GW00771872

MOSS AND LICHEN

Reaktion's Botanical series is the first of its kind, integrating horticultural and botanical writing with a broader account of the cultural and social impact of trees, plants and flowers.

Published

Apple Marcia Reiss
Ash Edward Parker
Bamboo Susanne Lucas
Berries Victoria Dickenson
Birch Anna Lewington
Cactus Dan Torre
Cannabis Chris Duvall
Carnation Twigs Way
Carnivorous Plants Dan Torre
Cherry Constance L. Kirker and Mary Newman
Chrysanthemum Twigs Way
Geranium Kasia Boddy
Grasses Stephen A. Harris
House Plants Mike Maunder
Lily Marcia Reiss
Moss and Lichen Elizabeth Lawson
Mulberry Peter Coles
Oak Peter Young
Orchid Dan Torre
Palm Fred Gray
Pine Laura Mason
Poppy Andrew Lack
Primrose Elizabeth Lawson
Rhododendron Richard Milne
Rose Catherine Horwood
Rowan Oliver Southall
Snowdrop Gail Harland
Sunflowers Stephen A. Harris
Tulip Celia Fisher
Weeds Nina Edwards
Willow Alison Syme
Yew Fred Hageneder

MOSS AND LICHEN

Elizabeth Lawson

REAKTION BOOKS

For mosses, lichens and their students, and the Lawson-Whalen-Fernandez-Hillman-Shelbred-Slattery-Rice family

Published by
REAKTION BOOKS LTD
Unit 32, Waterside
44–48 Wharf Road
London N1 7UX, UK
www.reaktionbooks.co.uk

First published 2024

Printed and bound in India by Replika Press Pvt. Ltd

A catalogue record for this book is available from the British Library

ISBN 978 1 78914 939 5

Contents

Introduction: The Cryptogamic Carpet 7

one Curious Vegetation *15*

two Moss: Versatile Minimalist *37*

three Lichen: Complex Individuality *67*

four Cosmopolitan Extremophiles *95*

five Bogland *117*

six Literary Ecology *137*

seven Curious Observers: A Field Trip *161*

eight #moss#lichen *187*

Timeline *215*

References *219*

Further Reading *243*

Associations and Websites *245*

Acknowledgements *247*

Photo Acknowledgements *249*

Index *251*

Introduction: The Cryptogamic Carpet

> Imagine a whole continent of naked rock, across which no covering mantle of green had been drawn – a continent without soil, for there were no land plants to aid in its formation and bind it to the rocks with their roots. Imagine a land of stone, a silent land, except for the sound of the rains and winds that swept across it. For there was no living voice, and no living thing moved over the surface of the rocks.
>
> RACHEL CARSON, *The Sea around Us* (1950)[1]

Of all the photosynthetic life forms that clothe the Earth and convert carbon dioxide to oxygen, mosses and lichens have earned first place as pioneers, creators and restorers of habitat. Despite morphological simplicity, their biological feats are extraordinary. A single moss cell isolated from a mother plant and carried by the wind can become a new plant in a new place, a phenomenon known as totipotency. A fragment of lichen, as small as a grain of pepper, carried on the claw of a small bird can start a new population. Moss and lichen make an odd couple – one evoking serenity and comfort in *Wizard of Oz* emerald-green cushions and golden-green feathery carpets, and the other evoking the uncanny in brilliant colours, fantastical forms and odd consistencies. Challenged by constant climate change over millennia, mosses and lichens speciated, becoming ever more adept at inhabiting every new nook and

Mosses thrive on waterfall spray in Rangárþing eystra, southern Iceland.

cranny of the Earth's unstable surface. They also offered shelter to ancient lineages of microinvertebrates, whose whisperings were the first living voices on Earth.

Linnaeus grouped the 'lower' plants, those that were unaccountably flowerless – algae, fungi, lichens, mosses, liverworts, hornworts and ferns – in the Cryptogamia, a name taken from the Greek for 'hidden' (Greek *kryptos*) and 'marriage' (Greek *gameein*, to marry). His sexual system of classification overturned centuries of belief in the chaste, feminine nature of plants.[2] He believed that a plant must have a flower, that the flower was the sole means of sexual reproduction and that all plants reproduced sexually. The apparent lack of sex in cryptogams was an impediment to acceptance of his theory.

Moss is a true member of the plant kingdom, while lichen belongs to no kingdom but its own. Indeed, the definition of a lichen keeps changing. Although Linnaeus classified lichens as plants, that changed in 1867 when Simon Schwendener revealed the lichen as a dual organism composed of one fungus and one alga. The dual conception of lichens launched the science of symbiosis, which gave rise to profound insights into how organisms successfully evolved and adapted to life on Earth. Recently other partners have been found. Lichens are now seen as ecosystems that behave like organisms.

The antlered jellyskin lichen (*Scytinium palmatum*) is rarely found without moss companions, Deception Pass, Washington State.

Lobaria pulmonaria, watercolour by Gherardo Cibo in Pietro Andrea Mattioli, *Discorsi on Dioscorides' De Materia Medica*, c. 1564–84. The iconic lichen is the focus of intense study today.

They are small, but they are everywhere, underfoot and overhead. Lichens cover approximately 7–8 per cent of the Earth's surface.[3] Ecologists estimate that 'in old-growth coniferous forests the dry mass of mosses and lichens rivals that of the foliage of all other plants.'[4] The alliterative terms 'cryptogamic carpet', 'cryptogamic cover' and 'cryptogamic crust' refer to mats of mosses, lichens and land-based algae, which cover 30 per cent of the Earth's ground and plant surfaces. A 2012 study involving chemists, geologists and biologists found that the cryptogamic carpet takes up as much carbon dioxide as that released per year by the burning of forests and other biomass, and accounts for nearly 50 per cent of the conversion of atmospheric nitrogen into forms usable by plants.[5] Biologists infer from observations of biocrusts and bacterial biofilms on adverse substrates such as deserts and volcanic rocks that mosses and lichens were among the

Cushion mosses and crustose lichens find community on rock slab, Highland County, Virginia.

Earth's first landscapers. Because the average existence of a species from origin to extinction is considered to be between 1 and 5 million years, establishing links between extinct versus extant species requires speculative detective work.

Evidence of the evolutionary history of mosses and lichens is found in fossils and through DNA sequencing techniques. Mosses and lichens do not fossilize well, and their habitats often defy preservation. The discovery in 1917 of moss- and lichen-like fossils in the Rhynie chert beds of Aberdeen, Scotland, set a start date of circa 410 mya for both, but some palaeobotanists consider these fossils 'ambiguous'. Recently, a credible peat-moss-like fossil found in carbonate deposits in Wisconsin indicated that the peat moss lineage began circa 607–460 mya.[6] Researchers using three different DNA sequencing methods produced age estimates of circa 602, 507 and 488 mya for moss, while geologists studying alluvial mudrock believe that the dramatic muddening of the planet circa 458 mya was caused by moss-like plants trapping sediment.[7]

Lichenization has been called 'a very ancient mode of fungal nutrition'.[8] The partners in the lichen symbiosis began existence well before moss appeared: cyanobacteria started oxygenating the planet around 2.7 billion years ago, fungi diverged from other life around 1.5 billion years ago and green algae evolved about 1 billion years ago. Fungi alone of the three is heterotrophic (*hetero*, Greek

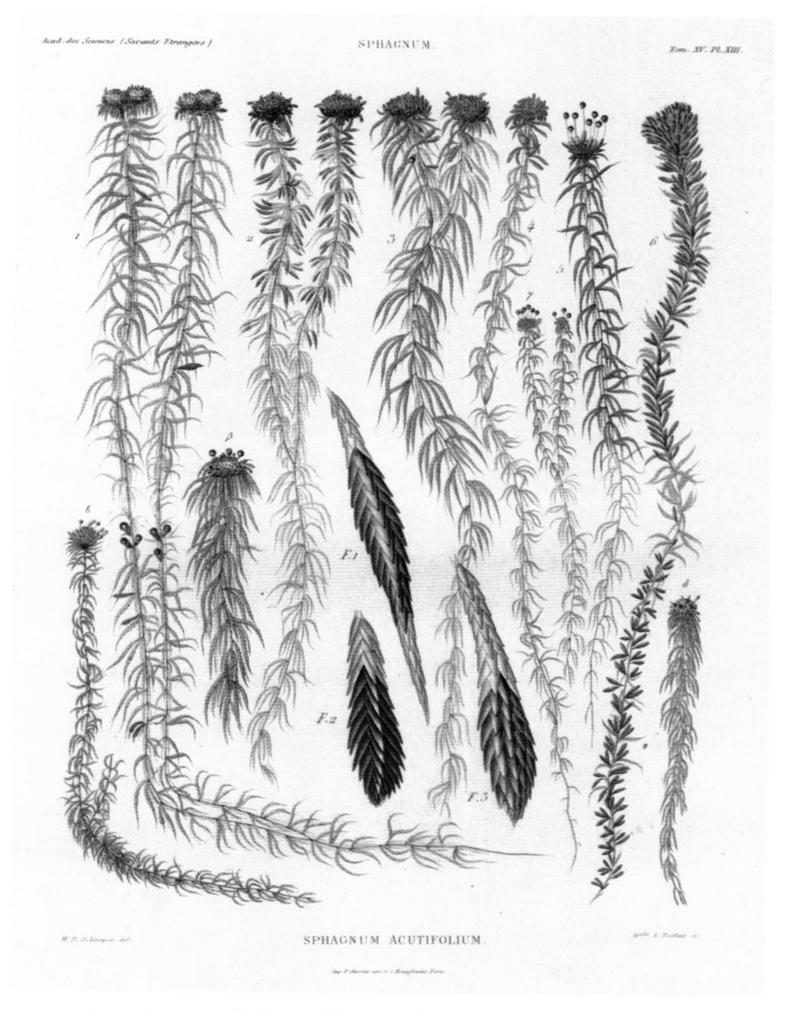

The unique morphology and beauty of sphagnum mosses portrayed by W. P. Schimper in *A History of Development of Peat Mosses* (1858).

Smoky-eye boulder lichen (*Porpidia albocaerulescens*) and moss share living space, Big Run State Park, Maryland.

for 'other' + *trophos*, Greek for 'feeder'), lacking the capacity to make food through photosynthesis. Symbiosis most likely began in the ocean, as fungi used these fellow organisms as sources of nutrition. The partnership then proved useful in the arduous process of terrestrialization as lichen acids became instrumental in breaking down rock. In 2005 the discovery of credible lichen-like fossils from the Neoproterozoic Doushantuo Formation of South China yielded an age estimate of circa 600 mya.[9] These specimens came from shallow seas rather than the land-based Rhynie chert fossils, supporting the notion that lichenization was an available lifestyle during terrestrialization. Data from a DNA technique called ancestral state reconstruction indicate that lichens have had a turbulent history: some fungi switched from the lichen lifestyle to that of decomposer or parasite and never relichenized. Others switched back, while still others swapped algal partners.[10] Another large DNA study of mutation rates in a group of extant lichens indicates a burst of lichen speciation circa 66 mya following the extinction of the dinosaurs

The common greenshield lichen (*Flavoparmelia caperata*) and moss find habitat on bark at the base of a tree, Highland County, Virginia.

and many higher plants, which gave the larger leafy/bushy lichens (macrolichens) room to diversify.[11]

Mosses and lichens are models of resilience and longevity. Mastering drought is their forte. Both can endure death-like desiccation for months and even years in a herbarium drawer and suddenly spring into vigorous regrowth upon watering. The name 'lichen' is said to be derived from the Greek verb *leichein*, 'to lick'. There are those who say that lichens are 'immortal'. The map lichen (*Rhizocarpon geographicum*) in Greenland, which grows just 1 centimetre every one hundred years, has been estimated to be 8,600 years old.[12] Life forms on the cusp of human visual perception, their appearance gives few hints as to how they so successfully persist in some of the harshest conditions on the planet. They carpet the Earth, but how without flowers and seeds? There is much to learn from these masters of refugia, past, present and to come.

Page from Johannes Dillenius's herbarium showing a specimen of the Great Goldilocks moss (*Polytrichum commune*), most likely collected by him.

one
Curious Vegetation

Before Dr. Dillenius gave me a hint of it, I took no particular notice of mosses, but looked upon them as a cow looks at a pair of new barn doors; yet now he is pleased to say, I have made a good progress in that branch of botany, which is really a very curious part of vegetation.

JOHN BARTRAM, Letter to Mark Catesby (1741)[1]

Since the use of the microscope has become familiar, we know that these dusts are worth paying attention to.

RENÉ-ANTOINE FERCHAULT DE RÉAUMUR, 'Observations sur la matière . . .' (1716)[2]

Close observation of mosses and lichens began in the seventeenth and eighteenth centuries. Like cryptogams themselves, the observers were a mixed bag – apothecaries, naturalists, physicians, botanists, working-class artisans and gentlemen (and undoubtedly gentlewomen) of leisure. Historian Hilda Grieve writes, 'In the 17th and 18th centuries it was a compliment to be proclaimed "a curious person". Curious persons were understood to be ingenious, skilful, patient and precise in research, connoisseurs in whatever brand of knowledge they pursued.'[3] The curious found the study of mosses and lichens compelling because they elicited wonder and close attention and posed problems, perfect specimens for the 'cabinets of curiosities' or 'wonder rooms' that became popular circa 1600.

Letters exchanged among early observers show that they often referred to each other as 'moss-croppers', mosses and lichens being grouped together at that time. Their voluminous correspondence has been called a form of 'epistolary science'.[4] Richard Richardson (1663–1741), an 'able and amiable' and prosperous gentleman doctor who lived in the heart of Yorkshire's textile district, had studied botany at Leyden University. From his correspondence, published by his great-granddaughter in 1835, we learn that he mentored a convivial group interested in cryptogams.[5] On 12 February 1701 William Vernon (1666–1711) wrote to Richardson from London: 'Mr. Sherard and I drank your health for three nights together: he is settling the mosses, and has set Mr. Bobart's right. Mine, Dr. Sherard's, and Mr. Doody's are all well; but the top of all the moss-croppers is Mr. Buddle, who is a great help to us.'[6] 'Settling' meant assigning names. Adam Buddle (1662–1715), a fervent amateur botanist, developed his own system of classification, a *methodus* for 'the more quick discovery of plants, [to] help to remember them better, and make them more regular and easy in the heads of young simplers [herb-gatherers]'.[7] Failure to publish his *methodus* diminished his access to fame, but a descendant amended that in 2008 with an essay titling him 'Moss-cropper extraordinaire'.[8] In 1698 the Royal Society sent Vernon to Maryland to collect plants. On his return he assisted John Ray with the cryptogamic sections of Ray's final volume of *Historia plantarum*.[9] In 1702 Vernon wrote to Richardson that 'the time of mosses is now in prime.'[10]

The composition of a 'historia' was a popular effort marking the accumulation of knowledge on a subject. One of the first was Theophrastus' *Historia plantarum* of circa 350–270 BCE. Johannes Jacob Dillenius (1687–1747) was the first to give mosses and lichens a 'historia' of their own. He published *Historia muscorum: A General History of Land and Water* in 1741. *Muscorum* is derived from the Latin *musci* (mosses). At this time 'mosses' included lichens and a few other 'lower' plants. Dillenius engraved the 1,000 figures and 85 'large Royal Copper Plates' for the *Historia* himself.[11] He had been brought from Germany to Oxford to be the first Sherardian Professor of

Botany at Oxford in 1734, a position Sherard had been grooming him for. Sherard wrote in a letter to Richardson, 'I had a letter yesterday from Dr. Dillenius of Giessen . . . He was recommended to me as a person very curious in mushrooms and mosses, as I perceive he is: he inclosed a moss he thought new and very curious.'[12]

Reading *Historia muscorum* is like taking a field trip into the eighteenth century. The six hundred entries under 'mosses' encompass a hotchpotch of algae, lichens, liverworts, true mosses and other lower plants in a variety of habitats. The descriptions of the specimens and their localities are helpful to biologists today because even though a moss or lichen may have disappeared from a particular locality for many years, startling reappearances occur. 'Great Goldilocks', now known as *Polytrichum commune*, or haircap moss, was found 'on Heath Bogs, Sussex'.[13] Lichens, not well named at this point and certainly not easy to describe, were given numbers. He found a specimen that he numbered 127, placed in Series v category 'Lichenoides' and described as 'grey cloudy leather, other side yellow'. Dillenius engraved an entire page of cup lichens, which he aptly called the Coralloides, because they resemble corals. Collectors sent him specimens from Sweden, North America, Jamaica and Patagonia.

One of Dillenius's far-flung collectors was Pennsylvanian Quaker plantsman John Bartram (1699–1777), a man whose curiosity knew no bounds. Bartram had absorbed much plant lore from his father, a Quaker farmer and herbalist, but had a moment of epiphany in a field:

> One day I was very busy holding my plow (for thee seeist that I am but a simple plowman) and being weary I ran under the shade of a tree to repose myself. I cast my eyes on a daisy. I plucked it mechanically and viewed it with more curiosity than common country farmers are wont to do; and observing therein many distinct parts, some perpendicular, some horizontal. What a shame said my mind, that thee shouldst have employed so many years in tilling the earth

John Bartram, portrayed by American illustrator Howard Pyle for *Harper's New Monthly Magazine* (February 1880).

> and destroying so many flowers and plants without being acquainted with their structure and uses! This seeming inspiration suddenly awakened my curiosity for these were not thoughts to which I had been accustomed.[14]

From that encounter he began a programme of intense self-education in botany. He borrowed books from Benjamin Franklin's public library, learned Latin to understand Linnaean names, mastered the use of a microscope and in 1730 laid out the first botanical garden in America at his farm, 6 kilometres from Philadelphia. Now known as Bartram's Garden, it is run by the John Bartram Association, whose mission is 'to create equitable relationships among people and nature through immersive, community-driven experiences that activate the Bartram legacy'.[15] He eventually became 'Botaniser Royal to America' for George III in 1765 and is credited with changing the face of British horticulture. Linnaeus called him 'the greatest natural botanist in the world', and Dillenius paid tribute by sending him a copy of *Historia muscorum*.

Both Dillenius and Bartram profited from the patronage of Lord Petre (1713–1742), whose fervent enthusiasm for horticulture, and deep pockets, endowed a vast collection of specimens from the New World. It was not easy to transport living plant material: the transatlantic voyages would last eighty days or more, and intermittent wars and salt spray endangered specimens.[16] Peter Collinson (1694–1768), son of a wealthy Quaker cloth merchant in London, acted as go-between for the lord and the ploughman and mentored Bartram in letters for 35 years. To his trade partners in Virginia he wrote: 'Don't be surprised if a downright plain countryman should turn up – you'll not look at the man, but his mind for my sake . . . He comes to visit your parts in search of curiosities.' They never met, but their friendship across the ocean is a bright note in the history of botanical and horticultural collaboration. Collinson wrote: 'Wee fellow Brothers of the Spade find it very necessary to share among us.'[17] Sharing also encompassed arguing about the nature of species and contrasting aesthetic values. Scholar Stephanie Volmer writes that the letters exchanged between Collinson and Bartram 'open a window into a transformative period in the history of perception and description of the natural world'.[18]

William R. Buck (New York Botanical Garden) and Elizabeth P. McLean (Academy of Natural Sciences, Philadelphia) studied the correspondence among Bartram, Collinson and Lord Petre with a special focus on the mosses from Bartram that Dillenius used for *Historia muscorum*. Collinson in 1739 asked Bartram 'to pray get some Mosses for him [Dillenius]. He is now Engraveing his Collection of Mosses in order to publish them.' Bartram wrote back in 1740: 'in my Journey to Menesinks on the Eastern Branch of Delaware att the foot of the paqualian Mountains . . . I have collected . . . about 60 sorts of mosses . . . pray if thee canst Conveniently let Dr. Dillenius see them it may be there may be some new sorts.'[19] Bartram did find mosses new to Dillenius, who rushed them into the *Historia*, making engravings of them at the last minute without preliminary sketches. Certainly Bartram had trained his eyes to look closely, quite unlike a cow staring at a barn door, hoping for her grain.

Bryological historian Mark Lawley, who has studied the social history of the early bryologists and lichenologists, found that many were textile merchants or weavers.[20] Weavers were practised at close examination, often looking at fibres with a hand lens. Edward Hobson (1782–1830), a weaver raised in poverty, offered *Musci Britannici* for sale in 1822. Few copies could be produced because Hobson placed actual dried specimens on each page by hand.[21] Richard Buxton (1786–1865), an apprentice bat maker (maker of small children's shoes), collected plants for the popular medicinal 'diet drinks' of the day and became an ardent self-taught botanist. He was so adept

Johann Hedwig named the common apple-moss (*Bartramia pomiformis*) after John Bartram.

with moss that William Hooker, director of Royal Botanic Gardens, Kew, wanted to hire him as an herbarium assistant. History has not forgotten the story of his transformation from an illiterate sixteen-year-old to the polished writer of *A Botanical Guide to the Flowering Plants, Ferns, Mosses and Algae Found Indigenous within Sixteen Miles of Manchester*.[22]

Moss-cropping may have begun in the field, but the work of identification and naming was largely carried on in pubs by working-class artisans. Historian Anne Secord has researched the culture of 'pub botany' in eighteenth-century Lancashire. Herbals, widely published in the sixteenth and seventeenth centuries, provided engravings and descriptions, and the Linnaean system of classification, which standardized naming, made plant identification and the naming of new species more accessible.[23] Secord writes that the pub became a place for memorizing and learning to pronounce Linnaean names. Gingham weaver John Horsefield (1792–1854), president of a number of Lancashire botanical groups, described the protocol followed by the presenter of a specimen:

> taking a specimen off the table . . . [he] gave it to the man on his left hand, telling him at the same time its generic and specific name; he passed it on to another, and so on round the room . . . But, from the noise and confusion caused by each person telling his neighbour the name of the specimen, some being unable to pronounce it, some garbling it, and all talking at once, we have been constrained of late years to adopt another method.[24]

The modified method involved silence as the president named the plant and passed it around. Patrons paid 'wet rent' to publicans to store herbals and other volumes for use as plant identification aids during society meetings at the pub. Horsefield mentored Hobson and Buxton, introducing them to others who liked botanizing in meres, mosses and moors. In December 1811, at Hobson's request, Horsefield dedicated a Sunday pub botanical meeting to mosses.

Though a specialist in higher plants, such as daffodils, Horsefield collected twenty different kinds of mosses. A few years later his discovery of a very small moss, *Weissia templetoni* (later renamed *Entosthodon templetoni*), would be pasted into Hobson's *Musci Britannici*.[25] Manchester historian John Percy writes, 'Most of our knowledge of the once rich floras of the surrounding mosses and moorlands come from men like these,' moss here meaning bog.[26] Secord makes clear that pub gatherings were not motivated by financial or social reward, but rather the attainment of knowledge. Horsefield wrote, 'we instruct one another by continually meeting together; so that the knowledge of one becomes many, and we make up for the deficiency of education by constant application to the subject.'[27]

At the same time as Bartram reached for the microscope to identify mosses in America, French naturalist René-Antoine Ferchault de Réaumur (1683–1757) was identifying lichens microscopically in the powders he scraped off stones and plaster.[28] Réaumer and others throughout Europe in the eighteenth century were interested in the many kinds of 'dusts' made more visible by the microscope – none more confusing perhaps than the 'dusts' produced by the cryptogams. In late winter and early spring, when mosses turn into pincushions, their green mounds 'bearded' with a forest of slender brown stalks topped with minute knobs, it is possible to wave a hand over them and see clouds of 'dust' emerge. These 'dusts' are much smaller than the seeds of flowering plants and appear empty inside. Since no other term or function was known for them, and Linnaeus had said that all plants produce seeds, the assumption was simply that they had to be seeds.

Scholars Lincoln and Lee Taiz write in *Flora Unveiled* (2017), 'Just as the telescope opened up the universe to human curiosity, the microscope revealed the formerly invisible world of the very small.'[29] The early microscopes were called fly glasses because of the popularity of viewing fly eyes. Galileo himself adapted his telescope into a compound microscope to magnify the eye of a fly. He and his colleagues called themselves the Academy of the Lynx-Eyed

Fringed hoar moss (*Hedwigia ciliata*): the intrepid, beleaguered natural history explorer Palisot de Beauvois (1752–1820) named an entire genus after Hedwig.

(Accademia dei Lincei). In 1663 Robert Hooke, demonstrator for the Royal Society of London, viewed moss 'leaves' under the microscope and named the honeycomb structures he saw 'cells'.[30] Although many microscope makers were at work throughout Europe, developing a scientific community whose members trusted one another would take time. At the end of the seventeenth century observers were still arguing over whether spontaneous generation of molluscs, insects and even stones was possible from the powders or dusts they sometimes called germs. Historian of microscopy Marc Ratcliff writes, 'The Italian programme on the seeds of cryptogams . . . agreed with antispontaneism, and was debated by nearly everyone in Europe.'[31] The fungi in particular were 'a headache for the Italians' because the spontaneists had the most plausible ideas thus far – that they

Engraving showing Johann Hedwig with microscope, hand lens and clump of moss, 1793.

arose by fermentation of plant lymph or through 'abnormal growth of their fibres'.[32]

Linnaeus had named the flowering plants phanerogams, those with 'visible marriages'. What could the cryptogams, those with 'hidden marriages', be up to? Transylvanian botanist Johann Hedwig (1730–1799) was able to identify the marriage partners, the marriage and the progeny with the microscope, becoming the 'Father of Bryology' in the process. Hedwig had begun armed with a 'meagre excerpt' of Dillenius's *Historia muscorum* and a 6× hand lens, but soon German botanist J.C.D. Schreber gave him a small library, and J. G. Köhler, inspector at the Mathematical-Physical Salon in Dresden who was known for drawings of lunar mountains, gave him a compound microscope made by Christian Friedrich Ernst Rheinthaler (*fl.* 1760–83) of Leipzig.[33] Hedwig was able to increase the magnification from 50× to 290×. Historians describe him as patient and conscientious, making repeated observations, postponing publication until he was sure. Some of the early microscopists shared unverifiable observations and inflated their data, exuberantly declaring magnifications of 2,600× to 8,000×, impossible for the instruments of the time.[34] His microscope had seven objectives, and he reported the number of the objective and its magnification on his figures. His drawings documented his microscope-related observations. His method of dissecting mosses with needles and small knives and then mounting fragments or sections on glass slides is still practised.

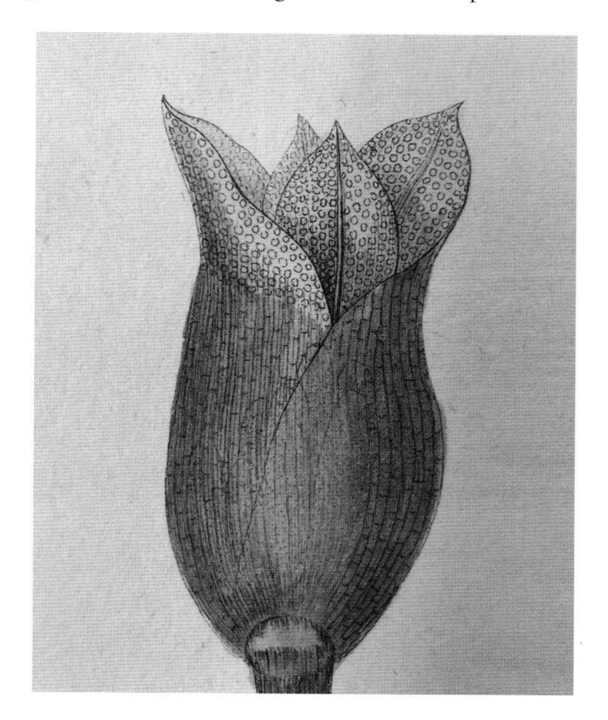

Hedwig's drawing of a moss bud (*Leersia ciliata*, later renamed *Encalypta ciliata*), from *Species muscorum frondosorum* (1801).

These methods allowed him to make observations 'of the true genitals of mosses and their reproduction by seed' (which he later termed spores).[35]

On 17 January 1774 he saw *Kugelchen*, 'little particles', swimming from minuscule clusters of tiny – even for mosses – leaves ('plantula mascula', later termed antheridia). Though he could not see the two flagella, he assumed that their motility indicated that they were sperm, already implicated in other contexts in sexual reproduction. Slightly larger clusters of leaves he termed female structures ('flos femineus', later termed archegonia). He saw microscopic green threads growing from the fine dust that blew in clouds from moss capsules. Hedwig's discoveries were underestimated or even misunderstood at the time because of higher-plant vocabulary problems. It was becoming somewhat evident that lower plants did not produce seeds but rather spores, which are not analogous. Hedwig had been convinced that there must be sexual reproduction in the mosses and lichens because he was an ardent follower of Linnaeus' system. However, his discovery of sperm in mosses did not end the great debate over the question of universality of sex in plants, a question the cryptogams had inspired:

> most classificatory schemes of that period . . . were based on the belief that all or most cryptogams were asexual . . . Those who argued for plant sexuality were assuming that the antheridia were male sex organs and that the spores were seeds; those who argued for the asexual nature of these plants were assuming that the spores were not seeds and that sex organs were entirely lacking. Both sides in the controversy therefore argued that the plants must be entirely sexual or entirely asexual.[36]

Nothing was more confusing than sperm – some biologists argued for the 'animality of algae' and 'ambiguous vegetables' that could transition into animals, based on observations of the release of swimming

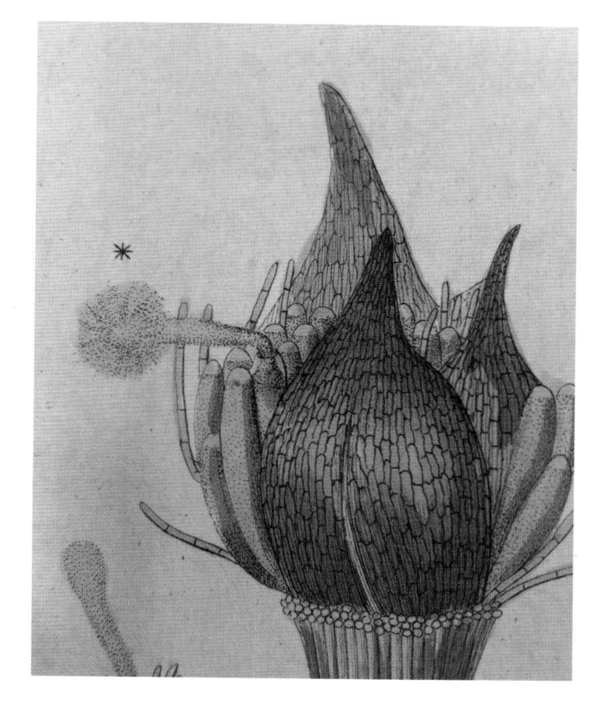

Hedwig's drawing of moss sperm being released from an antheridium of *Encalypta ciliata*, from *Species muscorum frondosorum* (1801).

sperm.[37] Discussion was polarizing: observers even accused their colleagues of 'faulty microscopy'.[38]

To view Hedwig's four-volume *Descriptio et adumbratio, microscopico-analytica muscorum frondosorum*, a work covering ten years (1787–97) of close observation and detailed drawing, is to see mosses as giants. In each plate Hedwig presents the moss in 'naturali magnitudine', life-size, usually 6–13 millimetres high. However, in the centre of each plate a larger-than-life version stands stunningly visible at 15–20 centimetres high. These illustrations have been called the most accurate and beautiful depictions of mosses of any time period.[39] No nuance of the minimalist architecture of mosses escaped Hedwig's eye: delicate twists of the top of the stalk (seta) supporting the capsule; the various curls and attenuations of the leaf tips; the many different arrangements of leaves, sometimes clasping, sometimes loose; areas of reddish and gold colouring particular to parts of the stem and capsules; sperm 'exploding' out of containers; striations of the capsule and peristome teeth; ribs of leaves; and lacy patterns of the cellular structure of leaves visible because most moss leaves are just one cell thick. He established the foundation of moss taxonomy and revealed the great diversity of mosses to a growing audience.

Hedwig became an authority on lichens as well. A writer in *The Edinburgh Encyclopaedia* of 1830 when discussing his contributions states, 'The fructification of lichens, or the mode in which they are propagated, has long been a botanical problem.'[40] At that time the strange appearance of lichens was thought to be indicative of their stressful lifestyle. The writer continues: 'we might naturally expect that lichens would, in that respect, be adapted to the contingencies

A photograph of *Encalypta ciliata*, the fire extinguisher moss, complements Hedwig's drawings, abandoned talc mine, Jeseniky Mountains, Czechia.

to which they are exposed, of being successively scorched, drenched, and frozen on the same barren rocks.'[41] Further, the writer articulated the mysterious ubiquity of lichens: 'It has often excited surprise that plants which so seldom present their parts of fructification, should yet be abundantly spread over our rocks and trees.'[42] Some were described as leprose, forming shapeless powdery masses, and others as crustose, forming scattered mosaics of hard granules, while the surface of leafy lichens showed many 'excrescences' – tiles, tubercles, shields, saucers, pustules. Some protuberances were termed 'fruit' or fruiting bodies. Hedwig unfortunately sought analogies with the flower in his efforts to prove Linnaeus right. His vertical sections showed 'a congeries of cells' that he thought might be pollen or a male flower. He also sectioned the cups often seen on lichens, which revealed strings of egg-shaped 'bodies'. He called them spores (a name still retained), considering the cups to be flower-like. Though unable to complete the story of lichens, Hedwig enhanced their reputation. He illustrated all the stages in the life cycle of mosses, and his introduction of 'spore' for moss 'dust' rather than 'seed' was a huge step forward in seeing cryptogams as other than primitive flowering plants.[43]

His contemporary Johann Wolfgang von Goethe (1749–1832) had a passionate interest in botany and is said to have admired Hedwig's 'impeccably organized' herbarium, now housed at the University of Geneva.[44] In 1776 Karl August, Grand Duke of Saxe-Weimar-Eisenach, lured Goethe into his service with the offer of a house and garden outside Weimar. There Goethe created flowerbeds for botanical observations, developing his theory that plants are just leaves. Visits to the Thuringian Forest inspired a special interest in the 'lower' plants, especially mosses and lichens. Goethe's thoughts about members of the plant kingdom from low to high reveal an evolutionary perspective:

> The ever-changing display of plant forms, which I have followed for so many years, awakens increasingly within

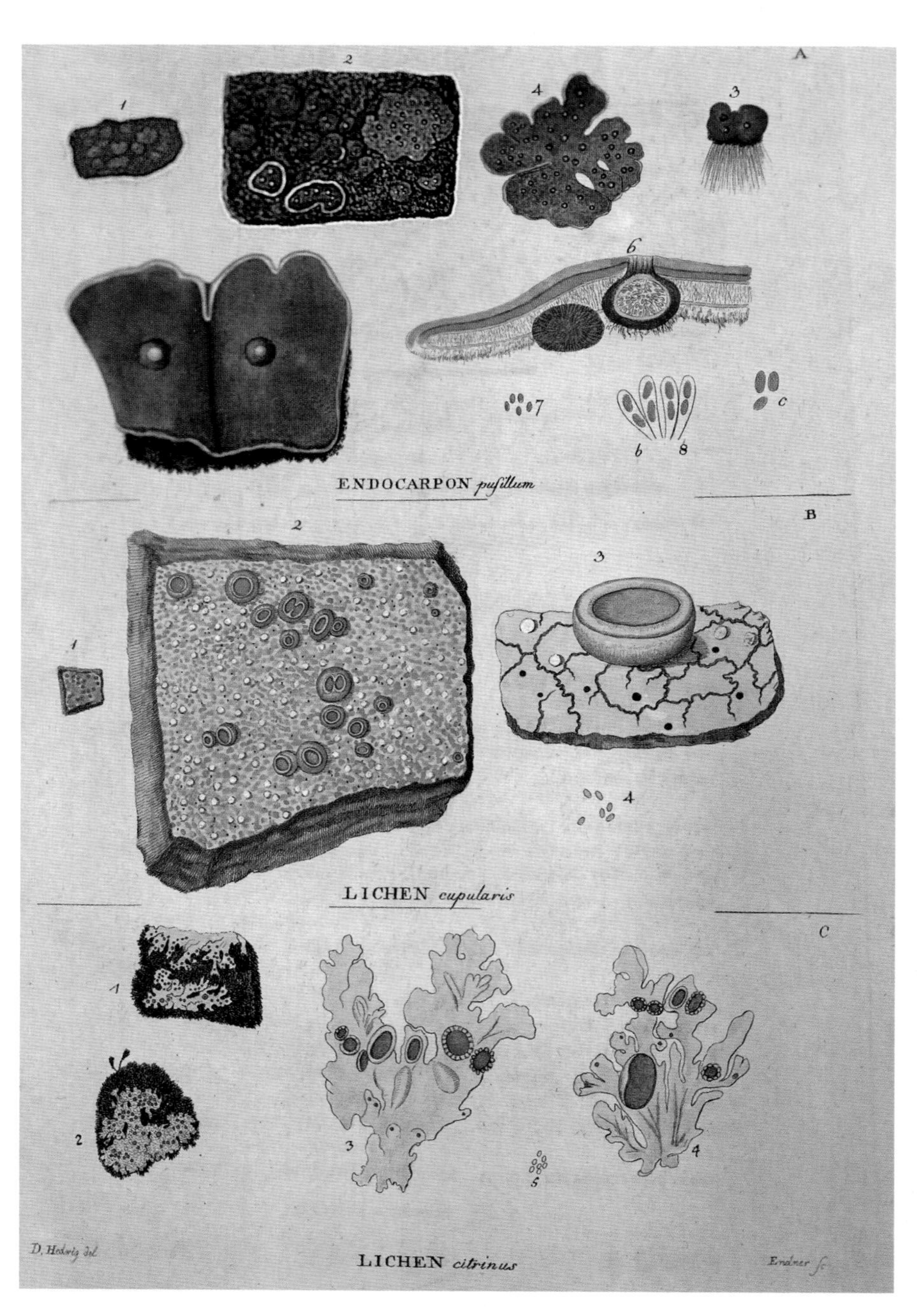

Hedwig's lichen drawings, from *Species muscorum frondosorum* (1801).

> me the notion: The plant forms which surround us were not all created at some given point in time and then locked into the given form, they have been given . . . a felicitous mobility and plasticity that allows them to grow and adapt themselves to many different conditions in many different places.[45]

Goethe published his botanical philosophy in *The Metamorphosis of Plants* in 1790.

Linnaeus' sexual system had been enthusiastically accepted after considerable dismay over the explicit mention of 'genitalia' in plants. However, his groupings were artificial in that he made no attempt to show relationships among his orders, which were based on numbers of stamens and pistils. By 1830 some would be ready to break free of Linnaeus' system. The spirit of the Enlightenment in France inspired a reassessment of Linnaeus' system, with proposals for a more naturalistic approach.[46] A clan of botanists in France led this movement. Antoine de Jussieu (1686–1758) made a fortune through his medical practice in Lyon, which he used to support the poor and finance his family members' botanical pursuits. He published *Plantae fungosae* in 1728, where he proposed a new class of plants to include lichens and fungi after observing their similar habitats and analysing them microscopically. He did not believe that they were simply 'diseased excrescences' of other plants.[47] Bernard de Jussieu (1699–1777) became the chief architect of the natural system. After earning several medical degrees, he became a teacher and theorist 'of genius' at the Jardin des Plantes in Paris. He determined relationships among groups of plants through the principle of multiple affinity, in which many characteristics of a plant in addition to the reproductive organs were considered, and carried on Antoine's work with fungi and lichens.[48] When Louis XV appointed him to supervise the royal garden at Trianon, Bernard used the opportunity to lay out part of the garden in beds ordered according to his system. After Marie Antoinette replaced Bernard's demonstration beds with an

English garden, his nephew Antoine-Laurent published Bernard's system from memory. Kliment Timiryazev (1843–1920) saw poetry in this living demonstration of a theoretical system:

> As if to justify its name, the natural system first saw the light not in the dust of a library nor the pages of a Latin folio, nor among the dried leaves of a hortus siccus, but living open to the sky and the spring sunshine, in the beds of the Trianon garden.[49]

Mosses and lichens must have found their way into his order beds, too small to be weeded out, but how the flowerless cryptogams fit into a classificatory lineage of plants was still a mystery.

Leipzig was also the setting for the last phase of uniting mosses with other plants. Wilhelm Hofmeister (1824–1877), a seller of sheet music, was the unlikely botanical star. Arthur Harry Church (1865–1937), inspiring Oxford lecturer and botanical illustrator, described Hofmeister's contribution:

> It was to Hofmeister, working as a young man, an amateur and an enthusiast, in the early morning hours of summer months, before business, at Leipzig in the years before 1851, that the vision first appeared of a common type of Life-Cycle, running through the Mosses and Ferns to Gymnosperms and Flowering Plants, linking the whole series into one scheme of reproduction and life-history.[50]

His was a vision of a particular sort. Being extremely myopic, he brought his eyes very close to specimens. Like Hedwig, he also became skilful in sectioning material for the microscope and drawing what he saw. Morton writes, 'His theoretical genius was firmly based on a consummate mastery of microscopical technique, unique powers of observation and interpretation, and an ability to delineate what he saw with impressive accuracy and beauty.'[51]

Wilhelm Hofmeister, like Hedwig, suffered disheartening mortality among his family members, but never ceased his prodigious efforts to visualize plant life cycles at the cellular level.

Hedwig had described the stages in the life history of the moss, but he failed to find something analogous in the ferns. Hofmeister was able to do that. By age 27 he had sectioned the reproductive areas of cryptogams and nineteen families of flowering plants and was able to document a common life cycle, often called an alternation of generations, because one is haploid and one is diploid. He drew chromosomes but did not know their significance. Tributes to Hofmeister's genius continue to appear.[52]

The true mystery was not what the cryptogams were hiding but rather what the higher plants were hiding – their gametophytes, reduced to only a few cells, remain within sporophytic tissues (cones and flowers), while the cryptogams display their gametophytes. Thus, the moss plant that we see is the gametophyte. When the slender

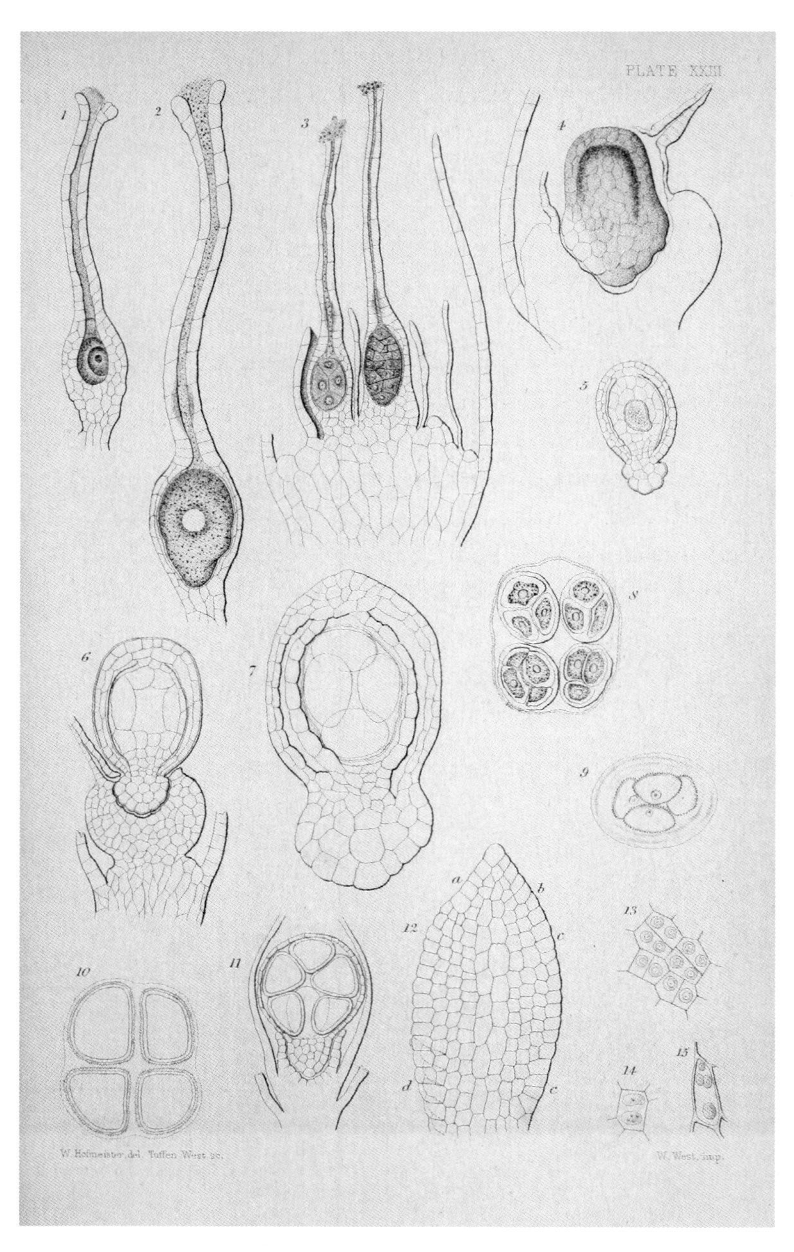

Hofmeister's drawing of fertilization and embryo development in the earth moss *Archidium phascoides* ('impregnation' is seen in no. 2), from *On the Germination, Development, and Fructification of the Higher Cryptogamia . . .* (1862).

stalk plus capsule, that is, the sporophyte, appear on top of moss tufts, both stages of the life cycle are visible, one atop the other. As botanist F. O. Bower would write, notions of 'lower' and 'higher' matter little in the workplace of life:

> on examination of the vegetation of any ordinary countryside, its uplands and lower levels, its swamps, streams, and pools, plants of the most varied affinity are found to be promiscuously shuffled together, and show little sign of ranking in their position according to their descent.[53]

Soon lichens would lose membership in the plant kingdom, but the ability of mosses and lichens to reproduce and carpet the Earth together make them the equal of flowering plants.

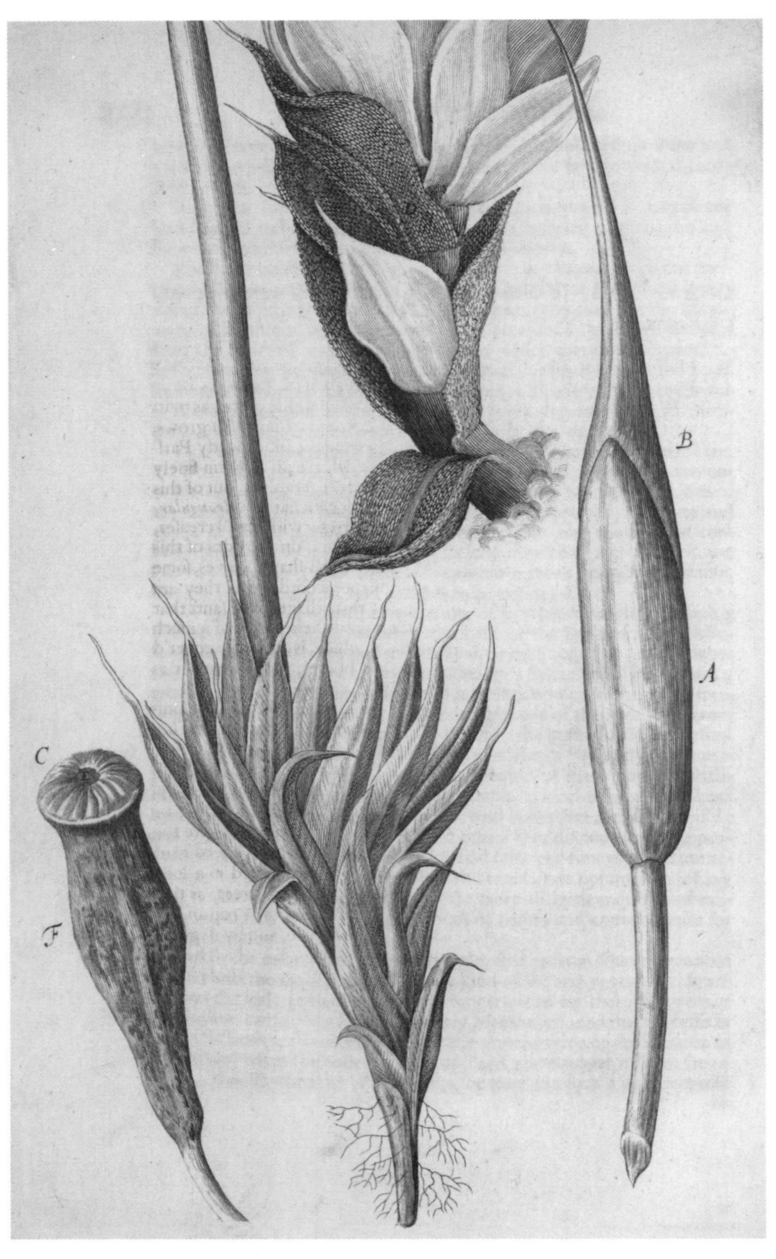

Robert Hooke's illustration of 'a common moss', observed through a microscope he had built himself, from *Micrographia* (1665).

two

Moss: Versatile Minimalist

Moss is a Plant, that the wisest of Kings thought neither unworthy his speculation, nor his Pen, and though amongst Plants it be in bulk one of the smallest, yet it is not the least considerable: For, as to its shape, it may compare for the beauty of it with any Plant that grows, and bears a much bigger breadth . . .

ROBERT HOOKE, *Micrographia* (1665)[1]

On 7 October 1663 Robert Hooke presented his 'microscopical observations' of the 'common moss' to the newly formed Royal Society of London. A brilliant mathematician as well as microscopist, he calculated that the average mass of one of the largest tropical trees in South America was about 2,985,984 million times bigger than the average moss.[2] Yet mosses have capitalized on their seeming deficits – in mass, stature, vascular tissue for transport of water and nutrients, two sets of chromosomes in the dominant phase, lignified cell walls and flowers – to become ecological giants in miniature. They have not grown bigger over evolutionary time because their minimalism succeeds beautifully.

Although previously described as evolutionarily stagnant, new research indicates that the reduced morphology of mosses belies a genetic dynamism that is greater than that of flowering plants.[3] Biologists have documented bursts of moss speciation following each of Earth's previous extinctions. One of the great students of mosses, British botanist M.C.F. Proctor (1929–2017), writes that bryophytes are not simply 'potential vascular plants' but rather they 'represent a radically different way of doing things'; he makes the following comparisons: 'Bryophytes . . . may be seen as the mobile phones, notebook computers and diverse other rechargeable battery-powered devices of the plant world – not direct competitors for the mains-based equivalents, but a lively and sophisticated complement to them.'[4] While it is easy to get lost in the details, they matter in getting to know mosses.

The Moss Life Cycle

Biology students the world over memorize the life cycle of the moss but rarely appreciate its importance in the history of land-plant evolution. It is often portrayed as a small circle inside a larger circle, populated by elfin figures propelled by arrows. The main goal of the diagram is to demonstrate that there is an alternation of generations: the gametophyte (haploid, carrying one set of chromosomes), which is green and long-lived; and the sporophyte (diploid, carrying two sets of chromosomes), which is non-green and ephemeral. Gametophytes produce gametes (sperm and egg cells), while sporophytes produce spores. Biologists still ponder the relative merits of whether an organism is haploid or diploid. Although having two sets of chromosomes has the advantage of masking harmful mutations, haploid organisms seem to be equally successful, even in the animal kingdom.

Spores mystified early researchers, who thought they must be seeds, despite their being so small and seemingly empty. However, they contain the genetic machinery for keeping the life cycle revolving and are decay-resistant, nutrient-rich, long-lived and produced in large numbers. Estimates range from 5.5 million spores per capsule for the tiny ephemeral bug-on-a-stick (*Buxbaumia aphylla*) to only 120,000 spores per capsule for the more substantial, ubiquitous fire moss (*Ceratodon purpureus*).[5] Researchers follow their presence in the aeolian dusts that circle the globe and in spore banks in the soil, testing the hypothesis that 'everything is everywhere but the

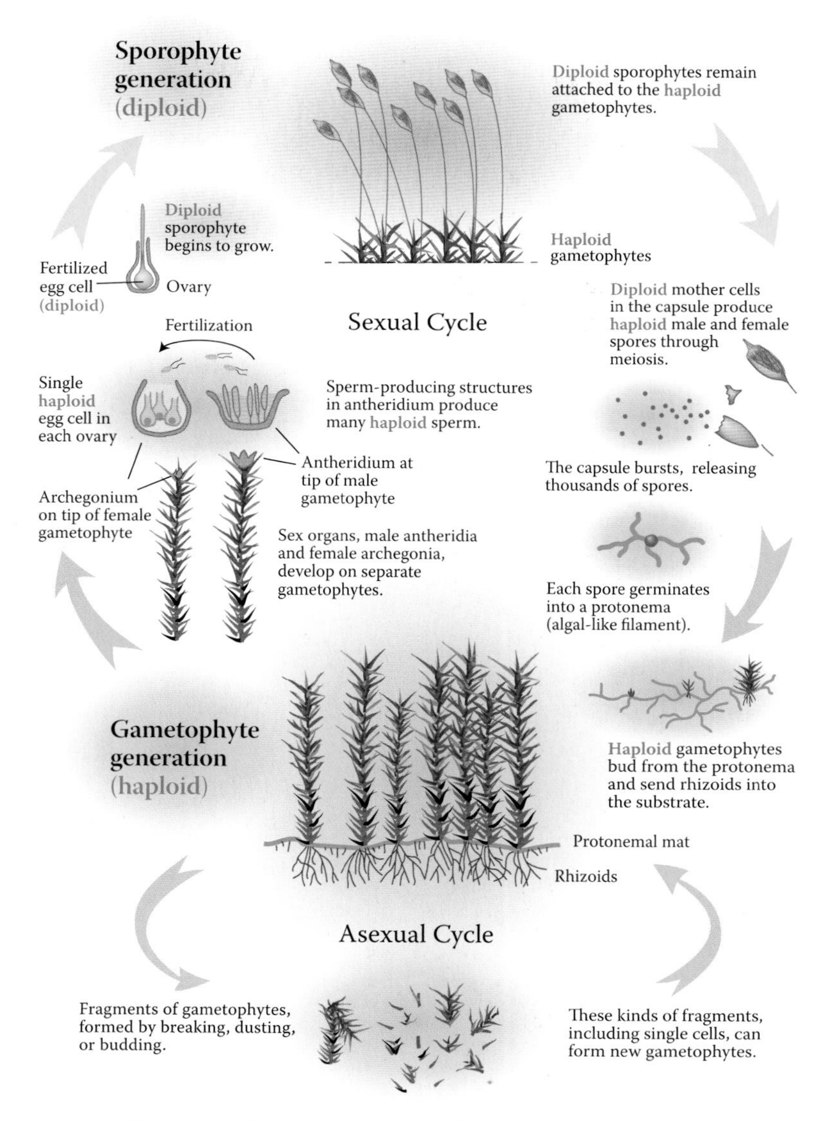

Deciphering the moss life cycle was a key advance in plant biology.

Overleaf: At least six species of acrocarp mosses inhabit this moss community found on the sandstone cliffs of Saunton Down, North Devon.

environment selects'.[6] The occurrence of disjunct populations is an interesting puzzle for bryologists. There are sixty mosses that occur at both poles but not in-between (bipolar distribution). Bryologist Lily Lewis developed a method for screening feathers for moss spores and fragments (diaspores) in the plumage of birds, such as the American golden plover and the red phalarope, which travel from Arctic breeding grounds to South American wintering sites. She found 23 bryophyte fragments, showing that bird plumage can transport diaspores over 4,830 kilometres or more.[7] Bryophyte consultant Des Callaghan reported in December 2022 the viability of spores of the endangered moss *Physcomitrium eurystomum*, which were unearthed at a 40-centimetre depth and dated at one hundred years old. Callaghan posted an image on X (formerly Twitter) of a glowing field of moss gametophytes grown from these spores.[8] He also reported that this moss completed its entire life cycle, from spore to spore, in just three months.

The Gametophyte

When a spore falls on a hospitable habitat, a filament of green cells emerges, the protonema (pl. protonemata) barely visible as a minute green web. Along its length, buds develop. Each bud on a protonemal strand grows into a single gametophyte. When the strand senesces, each gametophyte becomes a separate entity. One exception to the insignificant nature of the juvenile filaments is the clandestine cave moss or luminous moss (*Schistostega pennata*). The only member of its genus and family (Schistostegaceae), it is also known as goblin gold and the 'emerald light bender'. The protonemata grow deep within caves and in tree-root cavities in the least amount of light that allows germination of the spores. The chloroplasts of the lens-shaped protonemal cells cluster together in response to pinpoints of light. Refraction of reflected light within the cells yields the effect of a sprinkling of luminescent emeralds in dark recesses. The 'romance' of seeing luminous moss in situ was beautifully described by Austrian botanist Anton Kerner von Marilaun (1831–1898) in *The Natural History of Plants* (1863):

> On looking into the interior of the cave, the background appears quite dark, and an ill-defined twilight only appears to fall from the centre on to the side walls; but on the level floor of the cave innumerable golden-green points of light sparkle and gleam . . . If we reach curiously into the depth of the grotto to snatch a specimen of the shining objects, and examine the prize in our hand under a bright light . . . there is nothing else but dull lustreless earth and damp, mouldering bits of stone of yellowish-grey colour. Only on looking closer will it be noticed that the soil and stones are studded and spun over with dull green dots and delicate threads, and that, moreover, there appears a delicate filigree of tiny moss-plants, [each] resembling a small arched feather stuck in the ground.[9]

The 'feathers' are the gametophytes.

Moss growth is modular. Each gametophyte is essentially a module produced by an apical cell, which is shaped like an inverted tetrahedron (a pyramid standing on its head). The three oblique triangular surfaces, described as 'cutting faces', generate derivatives or daughter cells in a clockwise sequence.[10] As the daughter cells divide, they yield three vertical rows of cells termed metamers. Groups of metamers form organizational units called modules. Researchers analyse each gametophyte as 'a hierarchy of modules'.[11] Branches occur where hormones pool in cells called initials, which then produce either buds or leaves. The apical cell of the bud determines the growth of the branch, while other initials produce leaves. Distinctive patterns, whose variations are determined by each species's genes, develop as these modules are repeated.

Shoot branching, a key step in the evolution of land plants, is thought to have evolved independently in mosses and flowering plants after their divergence from a common ancestor, in an example of parallel evolution.[12] While both groups have the same three 'ancient hormones' governing branch architecture, the fact that flowering plants did not evolve from mosses makes the study of bryophyte development, particularly branching patterns, especially interesting.[13] Biologists recently analysed the architectures of 175 mosses to understand the kinds of developmental innovations and ecological factors, such as epiphytic versus terrestrial substrates, that led to the impressive structural diversity of mosses.[14] While the modular system is quite orderly in theory, for mosses in the field, and even in the laboratory, growth patterns become disarranged or rearranged by means that are not completely understood.

The gametophyte is a multipurpose, miniature widget that handles water absorption, retention and loss, produces sex cells and can

This bristle moss (*Orthotrichum columbicum*) recently attained specieshood after genetic analysis separated it, along with three other species, from a cryptic species complex.

The curved leaves of broom forkmoss (*Dicranum scoparium*) intertwine when dry, a means of conserving moisture.

disassemble for asexual reproduction, all without the vascularized roots, stems and leaves found in higher plants. The moss 'leaf' is usually just a single layer of cells about one-hundredth of a millimetre thick, 5–10 micrometres wide and 5–80 micrometres long.[15] Rhizoids are minuscule threads of cells that anchor the gametophyte to a substrate. They can also appear on stems, making certain mosses appear fuzzy or woolly (tomentose). Their biological efficacy in trapping sediment is significant, and they provide capillary spaces that move and retain water. Space does not permit description of all the minute bits and pieces (tomentum, gemmae, brood bodies, comal tufts, paraphyllia, paraphyses and so on) that represent fine adjustments to the moss 'business plan', especially useful as a means of asexual reproduction.[16] Some bryophiles advise petting mosses to release fragments for dispersal.[17]

To continue Proctor's comparison of mosses to mobile phones and computers, one might compare the moss leaf to a microchip etched with 450 million years of Earth experience. Similarly, it functions in sophisticated ways in a small area. Bryologist James Shevock writes that the 'design of a moss leaf . . . is simply a remarkable feature of architecture'.[18] Every part of the moss leaf shows an infinite range of variation – in the apex, the costa (midrib), margin, overall shape and leaf base. The costa may extend beyond the apex, just to the apex, two-thirds of the way, one-quarter of the way, form a forked 'v' at the base, be very wide the length of the leaf, be lacking entirely – or there may be two midribs. The leaf margin may be smooth, saw-toothed, rolled, incurved, recurved, bordered with distinctive cells that may be green or clear, and more. Extra-long leaf tips (sometimes called awns or hairpoints), whether the result of the midrib extending or an extension of the leaf blade itself, are often distinctive, flourishing dramatically while serving useful ecological purposes such as reflecting light and creating a rough boundary layer to slow down air movement and conserve water. Well-known churchyard species such as grey-cushioned grimmia (*Grimmia pulvinata*) and wall-screw moss (*Tortula muralis*) have outsized silvery hairpoints.[19] Cushions of grey-cushioned grimmia, whose hairpoints intertwine and curl down over the moss cushion when dry, have been compared to little grey mice, though the common name is hedgehog moss.[20]

Describing a moss leaf with even approximate accuracy is a daunting task. Many interesting adjectives have been thrown at the problem, for example, complanate (flattened in one plane), crispate (drying in a disorderly manner), falcate (curved like a sickle), lanceolate and ligulate (shaped like a strap), lunate, orbicular, obovate, oblong and mucronate (ending in a sharp point), muricate (having many sharp points), pulvinate (shaped like a cushion) and many more. Specialized adjectives function as useful shorthand. For example, spathulate is more concise than 'having a broad, rounded end'. When single adjectives fail to pinpoint the infinite spectrum of nuances, they are linked in strings of adjectives modifying adjectives, for example, deltoid cucullate, circinate falcate and ovate acute.[21] The cells of the leaf proper come in different patterns and shapes. Shevock compares the leaf cell arrangements of some mosses to stained glass windows and the surface ornamentation of their cell walls to small volcanoes.[22] Alar cells are two large, thin-walled cells, one at each corner of the base of the leaf. They swell and shrink quickly, allowing leaves to spread out or hug the stem to conserve water or receive more light for carbon gain.

Mosses are design specialists, creating spectacular diversity on a small scale. Although their leaves spiral around the stem in three ranks, each leaf being one-third of the circumference of the stem higher than the leaf below, they do so in different arrangements in a very small space, so innumerable special effects can be achieved. For example, consider *Rosulabryum subtomentosum*. It apparently has no common name, but its Latin name means 'little rose moss'; its leaves are gathered in bunches to form a rosette. The costa extends beyond the tip and bends backwards.[23] One of the difficulties of identification is that a moss when wet can look completely different when it is dry. Written descriptions complement the sometimes misleading clarity of photographic images by attempting to summarize the kinds of variation seen in the field. A look at various descriptions found in *Common Mosses of the Northeast and Appalachians* indicates the scope of the problem. The group of stringy mosses is especially hard to describe. The tangled thread moss (*Hygroamblystegium varium*) defies neat categorization:

> Appearance: Scraggly, creeping moss forming patches composed of a jumble of green, irregularly branched, fuzzy-looking threads . . . It bears the epithet varium for good reason: leaves even on the same plant can vary depending on water or nutrient availability. Plants range from blackened and bristly to pale yellow green and soft. Leaves are very variable. Lance to egg-shaped, 1–2 mm long, coming to a

short or long point, blunt or sharp. Midrib runs at least to midleaf, sometimes extending beyond the leaf as a short point, sometimes curving gently above midleaf.[24]

As one can see, it is necessary to register many details when getting acquainted with even just one moss.

A great step forward in identification of mosses for bryologists and amateurs alike occurred in 2020 when the donor-funded Northern Forest Atlas Foundation published *Mosses of the Northern Forest: A Photographic Guide*. It offers high-definition colour images of over 350 mosses in the Northern Forest Region of North America. Author and photographer Jerry Jenkins spent three years developing a digital photographic technique called 'focus stacking'. The method is painstaking and involves taking as many as two hundred images of

The mouse-tail moss (*Isothecium myosuroides*), also called the cat-tail moss, grows on tree branches and trunks in coastal rainforests of North America and Western Europe.

a single specimen while 'advancing the camera as little as 0.006 mm, roughly the width of a red blood cell'.[25] His colleague Sue Williams found specimens, identified them and did ecological surveys. Even Jenkins describes the technique as 'magical', and one is tempted to consider him a worthy successor to Hedwig's groundbreaking efforts in the eighteenth century to make the minute visible. It is a dramatic work of art, as is Michael Lüth's three-volume *Mosses of Europe: A Photographic Flora* (2019). Lüth has documented the landscapes, habitats and morphologies of mosses the world over in travels with bryologists.[26] Likewise, Neil Bell has made mosses more visible in *The Hidden World of Mosses* (2023) with the photography of Des Callaghan.

The Sporophyte

The mandate of the moss sporophyte is sexual reproduction. While fragments of a gametophyte can break off and start a new population, that colony will be a clone, genetically identical to the initial fragment. In mosses, sexual reproduction requires that a sperm cell find an egg cell. Mosses have swimming sperm, unlike flowering plants, in which non-motile sperm digest their way enzymatically down pollen tubes to locate egg cells following pollination. Melting snows and spring rains create the watery conditions conducive to the swimming of moss sperm, whose morphology has been described as bizarre, its two flagella likened to 'two coiled tails attached Cleopatra-style to a snake arm band'.[27] Their ability to swim any appreciable distance has been considered limited, 10 centimetres at the most, though raindrops can splash them further, especially in species of haircap mosses (*Polytrichum*), where the male reproductive structures (antheridia) occur in sky-facing 'splash cups'. New research, however, suggests heretofore unknown capabilities of moss sperm, such as desiccation tolerance, the use of a fertilization partner and long-distance travel. In 2012 a study using the fire moss *Ceratodon purpureus* found that fertilization in the laboratory occurred with the addition of rainwater or springtails, but rainwater plus springtails yielded twice as

Raindrops propel moss sperm from the red splash cups of *Polytrichum juniperinum* as far as 60 centimetres.

Pendulous moss formations deter erosion on cliffs, Shindagin Hollow, New York.

Springtails, shown here on a moss leaf, facilitate moss fertilization (*Dicyrtomina ornata*, Collembola).

much sexual reproduction as either alone.[28] They also found that the female reproductive structures (archegonia) emit a chemical scent that attracts springtails. In the Pacific Northwest there are approximately 300,000 springtails per square metre of forest, and they live predominantly in moss communities. It is thought that sperm adhere to the legs of springtails as they jump among the gametophytes and reach egg cells expeditiously as the springtails follow the sweet scent of the archegonia.

More than 50 per cent of mosses have separate male and female gametophytes. This makes water-dependent swimming sperm an even greater liability. Sperm from a male gametophyte face the daunting task of swimming over difficult terrain to a female gametophyte. A solution to this problem is the development of minute dwarf males (nannandry) from spores that happen to land on a female gametophyte. The dwarf males are just a few millimetres long and situate themselves conveniently right next to female sexual branches. The development of dwarf males as opposed to normal males is usually facultative. If spores land on a non-moss substrate, a normal male will be produced. There is evidence of hormonal control by the female as well. Female and male gametophytes are in direct competition for resources, so the development of dwarf males indicates that females are co-opting resources and facilitating fertilization by shortening travel distance for sperm.[29]

Sex may occur in mosses more often than is hypothesized. Genetic analysis of clumps of a haircap moss (*Polytrichum formosum*) in a beech forest at Kniphorstbos in the Netherlands revealed a surprising amount of genetic diversity. The assumption had been that clumps are largely clonal (genetically identical) because lack of water and slow sperm limit sexual reproduction. However, eleven genotypes were found in one of the larger clumps of about 40 centimetres across, and the entire population of clumps contained 26 genotypes. Paternity analysis revealed that one male parent came from as far as 2 metres away, but in some cases the male parent could not be found in the vicinity at all.[30]

After fertilization, tidy cushions of moss begin to look like dressmakers' pincushions bristling with pins, that is, the sporophytes. The diploid sporophyte embeds a foot in the gametophyte, receiving nourishment from the mother plantlet – a situation that is not technically considered parasitism because gametophyte and sporophyte belong to the same species. A wiry/springy stalk (seta) elevates the capsule or urn, which nourishes and disseminates haploid spores. When sporophytes are young, a cap of protective maternal tissue, the calyptra, covers the developing capsule.[31] As the seta or stalk elevates, the calyptra tears, sometimes creating a hairy cap or veil, which is a feature of the common haircap moss (*Polytrichum commune*).

Capsules display some of the most beautiful features of mosses. When the spores are ripe, a ring of cells called the annulus near the top of the capsule matures, allowing the lid (operculum) that seals the capsule to fall off, revealing a ring, or two, of teeth (peristome) surrounding the mouth of the capsule. The teeth always occur in multiples of four, with as many as 64 teeth in total. Depending on the level of humidity, they oscillate, ushering the spores in and out in a gradual fashion. A. J. Grout describes the beautiful peristomes of one of the bristle mosses, *Orthotrichum callistomum*, and the lurid cupola-moss (*Cinclidium stygium*) this way: 'These last two are so curious yet so beautifully adapted for their work that it seems almost like a fairy tale, and would be scarcely credible if told of some unknown tropical plant instead of having been seen and described by several of the most matter-of-fact botanists.'[32] He describes the peristome of *Polytrichum commune*, which has 64 short teeth attached to a trampoline-like membrane, as 'a most effective and ingenious pepper-box, entirely automatic in action'.[33] The wall-screw moss is named for the left-hand spiralling peristome teeth of the capsule.[34] Hedwig's stunning drawings of the details of the peristome teeth proved groundbreaking for classification, but also made apparent the rarely seen beauty of moss capsules. Like a flower, the moss capsule has many astonishing details that exemplify nature's investment in reproduction.

Sporophytes of the yellow moosedung moss (*Splachnum luteum*) have parasol-like capsules held aloft on elongated setae, Finland.

Of all the curious capsules, none are more so than the extraordinary lopsided capsules of the twelve species in the genus *Buxbaumia*, variously known as bugs-on-a-stick, humpbacked elves, elf-cap mosses and Aladdin's lamp mosses.[35] The first recorded sighting of *Buxbaumia aphylla* was reputedly in 1712, at the mouth of the Volga River, by Johann Christian Buxbaum (1693–1730), a German physician-botanist who travelled widely in Russia and Turkey in pursuit of natural history at the behest of Peter the Great. 'Aphylla' means 'without leaves', and aptly describes these species of ephemeral (annual) mosses in which the haploid generation remains juvenile, never moving beyond the protonemal stage – that is, there is no leafy gametophyte. However, the comparatively outsized capsules are visible to someone who can discern 'a lentil set on a toothpick' about 4–11 millimetres off the ground.[36] They are tilted at a 45–90-degree angle, with the flattened sides of the capsules facing the sun, an orientation said to enhance light capture and the raindrop dispersal of the spores. Like many of the early naturalists, Buxbaum died young, but Hedwig commemorated his eye for detail by naming the genus after him. Individuals may disappear for years before reappearing, which makes tracking their population biology for conservation purposes difficult.

In 2021 bryologists in France had a new insight. More focused observation had revealed that the 'allegedly undetectable' whitish-green protonemal mats often betrayed their presence with golden-brown gemmae, which are minute button-like clusters of cells that form asexual units of reproduction. Armed with this method of recognition, bryologists scoured the forests of France for the presence/absence of both stages, the immature gametophyte with gemmae and the curious sporophyte of the green bug moss (*Buxbaumia viridis*). The map of France became dotted with data points showing their results – *B. viridis* was found in 79 forests in 12 departments. Further, with its apparent invisibility unmasked, some fascinating behaviour was revealed. It appears that the gametophyte and the sporophyte of the green bug moss are rarely found together; their ecological preferences/requirements differ – the gametophyte prefers/requires rotting deadwood in shaded forest settings, and the sporophyte prefers frequently disturbed habitats with greater light.[37] Apparently this gametophyte–sporophyte 'quarrel' has not interfered with spread of the species. Asexual reproduction is an especially handy tool in cases like this.

The parasol-like capsules of the dung mosses, species in the family Splachnaceae, appeal in a different way. Fortunate observers in the field will be startled and amazed to see a crowd of ravishingly beautiful petticoats or parasols in shades of yellow, red and black presiding gracefully over cow pats, moose dung, sheep carcasses, owl pellets, antlers and the like, all the while emitting their own malodorous chemicals to attract flies (Diptera). The dung mosses, which rely on insects for dispersal (entomophily), find suitable habitat on faeces and carrion in bogs and meadows (coprophily). The flower-like parasols are modified portions of the non-spore-bearing lower portions of the capsule (apophyses) that have become modified to attract visitors. Not all species are showy. The cruet collar-moss (*Splachnum ampullaceum*) does not have a showy petticoat, but the swelling of the apophysis is still wide enough to afford a perch for a fly. It grows only in bogs on white-tailed deer dung. Robin Wall Kimmerer writes in 'Portrait of Splachnum' (2003) that 'the set of circumstances that converge to bring *Splachnum* into the world is highly improbable.'[38] While most mosses release clouds of spores adapted for wind dispersal, genera in the Splachnaceae produce small, sticky spores that lodge readily in hairy insect feet. Some, such as the yellow moosedung moss (*S. luteum*), grow only on herbivore dung, and others only on carnivore dung, while *Tetraplodon fuegiensis* grows only on the dung and carcasses of foxes. In 2019 *Tayloria octoblepharum* was found growing on tree shrew faeces that regularly fall into the 'toilet pitchers' of two species of pitcher plants in Borneo.[39]

This manner of attracting spore dispersers (entomophilous coprophily) is thought to have evolved multiple times in the approximately 73 species in the Splachnaceae. A team of researchers

'Bugs-on-a-stick': the capsules of *Buxbaumia aphylla* sporophytes rise above microscopic gametophytes.

analysed the chemistry of decay-mimicking odours in various species of *Splachnum*.[40] One of the least showy, the pinkstink dung moss (*Splachnum sphaericum*), produces more than fifty volatiles, including those indicative of fermenting sugar, flowers, herbivore faeces and, remarkably, moose urine. This species particularly likes trampled sheep dung found in wet meadows above a 360-metre altitude.[41] While most mosses are thought to have lost stomates during the

Capsules of the red parasol moss (*Splachnum rubrum*) rise out of boggy waters in Finland.

course of evolution, dung mosses release these chemicals via a few enlarged stomates on the sporophyte. In Norway the popular press takes note of the proliferation of the black-fruited stink moss (*Tetraplodon mnioides*), also known as the slender cruet dung moss and the fuzzy poop moss, on lemming carcasses in headlines such as 'Lemmings' Loss Is Bounty for Mosses'.[42] When lemmings die in large numbers after 'a lemming year' in barren and high-altitude northern areas, stink mosses proliferate. Dung mosses are thought to have evolved to capture nitrogen and phosphorus in nutrient-poor areas, nature's ingenious recycling at work.

While lemmings may never be in short supply, the fate of the lovely red parasol moss (*Splachnum rubrum*) is tied to the imperilled fortunes of the moose (*Alces americanus*) in North America.[43] In the last decade moose populations in Minnesota and New Hampshire have decreased by 60 per cent and 40 per cent, respectively, as higher temperatures lead to overheating and greater parasite loads (especially ticks), both of which lower birth rates. *Splachnum rubrum* grows exclusively on old moose dung. However, there is at least one study of *in vitro* cultivation of a dung moss, the cruet collar-moss.[44] Fortunately the beauty of *S. rubrum* has been captured by Michael Lüth. A browse through his online portfolio, titled 'Splachnum rubrum 2009', reveals the improbable beauty of these improbable mosses.

Acrocarps, Pleurocarps and Sphagnum

Making the leap from moss guidebooks to field observation can be unsettling. A drawing or photograph of a single gametophyte in a guidebook may give a very different impression than when gametophytes are massed together in the field. Ecologist Karl Mägdefrau (1907–1999) emphasized that 'Unlike most of the higher plants, bryophytes are not found as single individuals but in groups of individuals which have characteristic features depending on their family, genus or species.'[45] Bryologists recognize three groups of mosses based on growth pattern – the acrocarp (Greek *akros*, extreme, topmost), an upright, tufted form; the pleurocarp (Greek *pleurá*, lateral), a creeping, flattened, branched form; and the sphagnum or peat moss. Typical acrocarp growth patterns include clumps, tufts, cushions, groves and mounds, while pleurocarps tend to form carpets, mats, tangles, turves and wefts. Sphagnum mosses are uniquely themselves, their mop-headed gametophytes tangled together for support in watery environments. Although the distinction between acrocarps and pleurocarps does not have a genetic basis, it is often the first step for an amateur interested in naming a moss or two.

While acrocarps are considered less advanced evolutionarily than pleurocarps, their tightly aggregated gametophytes give them many advantages: architectural support, desiccation and pollution tolerance, facilitation of sexual reproduction and ease of interplant communication. Many acrocarps have worldwide (cosmopolitan) distributions. In fact, bryologist Brent Mishler describes the clump as a super-organism and compares it to a social organism like the beehive.[46] Jerry Jenkins of the Northern Forest Atlas Project calls clump-forming acrocarps 'supermosses'; his description of one, the fire moss, *Ceratodon purpureus*, illustrates how a clump becomes complex over time:

> In the open, the colonies are rounded, with the living stems sitting on top of a mound of old stems, rhizoids, pebbles that they grew up through, and fine soil that they captured. This is their biomass hoard; it stores nutrients and water; the living plants keep it from blowing or washing away and provide a rough surface that reduces airflow in the boundary layer and decreases evaporation. The first eco-morphic rule for extremophiles is: grow low and tight, stash biomass, take it with you when you go. From this point of view, this is less a plant growing on a rock than a self-contained ecosystem, built by the plant, that happens to be perched on a rock.[47]

Clumps tend to tumble and 'grow wherever they end up'. The ability to become transient at a moment's notice is an advantage. Clumps can increase their girth, dominating their habitat. In Alaska and the high Arctic the woolly fringe-moss (*Racomitrium lanuginosum*) often grows in such large colonies that it can be identified from a mile away. So too on the lava flows of Iceland. Images of gametophytes growing from the top up while still attached to dead layers beneath show how moss clumps gain stature over time.[48]

Acrocarps do well in the disturbed areas that result from human-dominated activity, occurring widely in inner cities and industrial areas, along the margins of major roadways, on railway lines and in unstable natural habitats, such as sand dunes.[49] Anyone who has walked on city streets has probably stepped on silver moss (*Bryum*

Tightly packed gametophytes fend off competition from higher plants, Grímsnes- og Grafningshreppur, Iceland.

Tufts of cushion mosses (acrocarps) gain height in layers as buds below the reproductive organs are released after fertilization to form the next layer of gametophytes (green), shown in *Tetraplodon mnioides* with most recent sporophytes above.

The brocade moss (*Hypnum imponens*), a feather moss native to North America, forms tightly knit wefts.

argenteum) densely filling the cracks of dusty pavements.[50] Also called silvery thread moss, it is described as possibly 'the most ubiquitous moss on Earth, but largely absent from its native undisturbed tropical habitat', and its biogeography has been studied extensively in journals as divergent as *Weed Science* and *Antarctic Science*.[51] Some of its spread is unwanted, especially on the putting greens of golf courses, a highly curated environment that it manages to intrude successfully.[52] In Antarctica the current population represents a persistence of potentially several million years' duration. It is a common candidate for green roofs because of its tolerance for high-light conditions and desiccation. Bryologist P. W. Richards, who studied Robert Hooke's drawing in *Micrographia*, believes that his 'common moss' of 1663 was a *Bryum* sp., plucked from a handy wall, whose descendants may still reside in London.[53]

The stiff, upright acrocarps act as pioneers for the graceful, prostrate pleurocarps, which spread laterally indefinitely, creating delicate wefts and colonial mats. Fifty per cent of all true mosses belong to a large group of pleurocarps, the feather mosses (Hypnales), with about 4,000 species. Many of these mosses branch pinnately like feathers and ferns. The dense wefts of the delicate fern moss (*Thuidium delicatulum*), which can be two to three times pinnate, suppress competitive vegetation. It was used in previous centuries to pad coffins and cradles but is now primarily sold commercially for use in terrariums. The electrified cat's tail moss (*Rhytidiadelphus triquetrus*) has been called 'untidy' in descriptions and owes its name to bristly tail-like shoot tips. The brocade moss (*Hypnum imponens*) is the opposite of untidy, its tightly packed leaves appearing neatly braided or combed. Feather mosses are iconic members of the forest floors of boreal regions, where they produce more biomass than the trees above and contribute significantly to the nitrogen pool through association with nitrogen-fixing cyanobacteria. The transfer of nitrogen from mosses to the soil of these nitrogen-limited ecosystems is said to follow a tortuous pathway, which requires further research.[54] The weft form is also useful in maintaining a clean, well-lit carpet. In the black spruce/feather moss climax forests of northern Alaska and Canada, fallen spruce needles are in effect woven into the weft as the moss grows over and around them. Even low-light-adapted mosses cannot survive dense cover.

Each moss tells a different story about life on Earth. After sufficient study, one internalizes a magnified image of mosses. Bryologists, who view mosses as Leviathans, would not sympathize with the disparaging statement of Herman Melville's narrator Ishmael during his ruminations on the classification of whales in *Moby Dick* (1851): 'It is

A self-supporting moss cushion sits aloft in a tree, Shindagin Hollow, New York.

by endless subdivisions based upon the most inconclusive differences, that some departments of natural history become so repellingly intricate.'[55] Instead, they see their intricate details as both ravishing and fundamental to their success in the workplace of life. Minimalists in size but not design and physiology, mosses merit attention for their profound impact on the Earth.

Cladina evansii (Evans's reindeer 'moss'), a fruticose lichen, seen here in the Naval Live Oaks Nature Preserve on the northwest coast of the Florida Panhandle.

three

Lichen: Complex Individuality

Gorgeous and weird, lichens have pushed the boundaries of our understanding of nature – and our way of studying it.
ED YONG, 'The Overlooked Organisms' (2019)[1]

Mosses fit tidily into the plant kingdom and in fact united the lower and higher plants, but that was not to be the case for lichens. They would never fit in. As Albert Schneider emphasized in his *A Guide to the Study of Lichens* (1898), 'The history of lichenology is a remarkable one. It indicates that this special science has progressed in a devious and interrupted course.'[2] Annie Lorrain Smith summed up the situation in her classic text *Lichens* (1921): 'Controversy about lichens never dies down.'[3] As if to prove her point, in 2019, science writer Ed Yong profiled lichens in *The Atlantic* in an essay titled 'The Overlooked Organisms That Keep Challenging Our Assumptions about Life'.[4] We learn it is best to approach lichens with an open mind, sceptical of definitions and generalizations. One lichenologist says that their way of being is 'contextual'; another notes that lichens challenge his ability to use language effectively.[5]

Their strange, scrappy lifestyle inspired bizarre theories as to their origin. Some suggested they were 'excrementitious matter, produced by the earth, the rocks or the trees', or perhaps 'the result of the decomposition of a higher vegetation' or 'pure precipitation from . . . vegetable juices'.[6] It was even proposed that they were 'transmutable

The pink earth lichen (*Dibaeis baeomyces*) spreads quickly over disturbed ground and has both crustose (the thalli) and fruticose (the pink apothecia) components.

into animalcules'.[7] Lichenology historian Charles Christian Plitt writes, 'It goes without saying, that most of these ideas were not founded upon direct observations, but were merely opinions.'[8] In 1967 the newly formed International Association for Lichenology advanced a formal definition of the lichen as 'an association of a fungus and a photosynthetic symbiont resulting in a stable vegetative body having a specific structure', a bland statement that failed to settle a Pandora's box of complex definitions and has prompted calls for revision.[9]

Taxonomists studying the external morphology of lichens found a wondrous and varied landscape – ruched, ruffled, puckered, jellied, spotted and dusted – that fuelled their passion for naming and classifying. In 1798 Erik Acharius (1757–1819), the last student to defend a dissertation in front of Linnaeus, compiled a description of all known lichens in Sweden, *Lichenographiae svecicae prodromus*. He also completed medical studies and eventually settled in lichen-rich Vadstena, a small town in Sweden, where he produced the bulk of his lichen work while occupying an official position as head physician at a new hospital for patients with venereal diseases.[10] He would name 330 species in 40 genera, becoming known as the 'Father of Lichenology', and influence Britain's William Borrer, who would earn the lesser but well-earned title of 'Father of British Lichenology'. Acharius's work taxonomizing lichens gave them credibility in the botanical community.

When microscopists looked at the internal structure of lichens, they found a confusing scenario. They saw filamentous, colourless (lacking in chlorophyll) threads and greenery, some in strings of blue-green pearls and others just 'little round green things' (LRGTs, the abbreviation favoured by students of algae). Sometimes there were discernible layers, with the LRGTs wrapped in nests by the threads. Observers assumed that they were dealing with a single organism. Clear, straightforward observations were at hand along the way, but hasty assumptions, misleading terminology, entrenched thinking and

Pelt lichens were put in shoes to deter the bites of rabid dogs (apple-pelt lichen, *Peltigera malacea*).

The ground-dwelling orange chocolate chip lichen (*Solorina crocea*), a tripartite lichen containing a fungus, green algae and cyanobacteria, produces a novel protein.

baseless assertions, combined with partisan rhetoric, led to years of misinterpretation and conflict in the biological community about the nature of the lichen.

An unfortunate misstep occurred circa 1825, when German physician Friedrich Wallroth (1792–1857), viewing sections of lichens under the microscope, decided to name the LRGTs 'gonidia' and to define them as reproductive bodies produced by the filamentous threads. He did see free-living LRGTs outside lichens but explained them as 'unfortunate brood-grains' (gonidia) that had failed to make a lichen thallus.[11] In 1833 Friedrich Traugott Kützing examined the same cells and wrote: 'Protococcus viridis [alga] is very often found on trees supporting the Parmelia [lichen]. If the structure of this lichen is examined under the microscope, the same spherules of Protococcus are found in the thallus, and, indeed, this Protococcus is fundamental to the existence of Parmelia parietina.'[12] Kützing was a knowledgeable phycologist, but his observation failed to dislodge Wallroth's faulty assumption.

In 1849 George Henry Kendrick Thwaites (1812–1882), expert microscopist and lecturer at the Bristol Medical School, investigated one of the jelly lichens (*Collema* sp.) under the microscope. When wet, they look like masses of black or dark-olive gooey, blistered bubbles; when dry, they create hard crusts that are important for preventing soil erosion.[13] He published a paper reporting his observation that the strings of pearls looked very much like *Nostoc*, a species of cyanobacteria (at the time called blue-green algae). Jelly lichens are now recognized as cyanolichens, those that contain cyanobacteria, rather than green algae, thus becoming capable of nitrogen fixation. His observations nearly told the whole story, and the history of lichenology might have proceeded in a more orderly fashion but for his almost immediate departure for Ceylon (now Sri Lanka) to become superintendent of the botanical garden at Peradeniya.

The jelly lichen also drew the attention of Heinrich Anton de Bary (1831–1888), an excellent microscopist famous for identification of pathogenic fungi such as rusts and smuts. In 1863 he essentially

The mourning phlegm lichen (*Lempholemma polyanthes*), a cyanolichen whose photobiont is the cyanobacterium *Gloeocapsa*, is usually found growing on moss over rock.

repeated Thwaites's observations and struck a note of incredulity that 'gonidia' could be other than algae: 'The gonidia correspond so strikingly in many ways to lower algae and particularly with regard to their reproduction, that it can be frankly stated that here Nature has brought a part of algal life into being for a second time.'[14] In 1866 de Bary proposed that gelatinous lichens such as *Collema* represented a case of parasitization of *Nostoc*.

An announcement clarifying the internal anatomy of lichens came at a meeting of the Swiss Natural History Society held in Rheinfelden, Switzerland, on 10 September 1867. Simon Schwendener (1829–1919) had been studying lichens in the lab of Carl Nägeli. He is quoted as telling his audience that day that 'in an entire group of lichens, gonidia and filaments are not in genetic connection, rather the latter are to be regarded as proliferations of fungal hyphae over algae.'[15] Fungi and algae living together! After ten years of studying lichens under the microscope it must have seemed a somewhat innocuous statement of the obvious to him, but it was a huge moment in the history of biology: it ushered in the science of symbiosis.

Schwendener assumed that the fungi were 'parasitic on algae' because that was the prevalent view of the nature of fungi at the time. He followed his talk with a series of publications in which he used an unfortunate metaphor for parasitism in lichens that disturbed experts throughout Europe and America. He described lichens as 'colonies' with the fungi acting as 'parasites, although with the wisdom of statesmen', and called the algae 'helotes' after the slaves of ancient Sparta, elaborating the scenario:

> This master is a fungus, of the class Ascomycetes, a parasite, accustomed to live upon others' work, its slaves are green algae, which it has gathered around itself, at any rate, holds on to and forces into service. It invests them as a spider her prey, with a fine meshed web, which gradually is converted into an impregnable integument, but, whilst the spider sucks out her prey and throws it aside when dead,

> the Fungus stimulates the Algae, found in its net, to more lively activity, in fact, causes them to grow larger and causes thereby a luxuriant growth and the thrifty appearance of the whole colony.[16]

Critics of Schwendener cited 'the vigor and longevity of lichens' as proof that parasitism could not explain this living arrangement.[17] Following the controversy from afar, Thwaites wrote in a letter of 1 October 1873 to the Royal Horticultural Society:

> But who has ever seen the gonidia of Lichens the worse for having the 'hypha' growing amongst them? These gonidia are always in the plumpest state, and with the freshest, healthiest colour possible. Cannot it enter into the heads of these most patient and excellent observers that a cryptogamic plant may have two kinds of tissue growing side by side without the necessity of one being parasitic upon the other?[18]

Thwaites deserves credit for having an open mind. The answer was no . . . or yes . . . or who knows? The battle over 'Schwendenerism' or 'the dual hypothesis' began. On the one side were the dualists, who embraced the idea of the lichen being a twosome, and on the other were the autonomists, who argued against this conception with near fury. To lichen taxonomists the dual hypothesis represented a demotion of the lichen, a profound change in status for both lichen and lichenologist.

The most vociferous autonomists were Finnish lichenologist William Nylander (1822–1899), leading lichen taxonomist of the day, and Reverend James M. Crombie (*c.* 1831–1906). Nylander had a reputation for being able to identify a lichen from anywhere in the world. Lichens and letters passed back and forth from Paris to Britain for 25 years as Crombie sought help in identifying specimens. Only Crombie's side of the correspondence has survived, unfortunately,

A foliose lichen, perhaps *Physcia subtilis*, colonizes a piece of marble discarded after a kitchen renovation.

but historians have made good use of it. Nylander led the battle. He treated Schwendenerism as a personal affront to lichens and by extension himself. He called the hypothesis 'an assertion of pure fantasy or slander'.[19] Nylander blasted dualists with 'vitriolic abuse', writing:

> From the biological point of view the Lichens are sufficiently differentiated by the indefinite longevity which characterizes them. The beautiful specimens of Umbilicaria pustulata in the forest of Fontainebleau are probably a little less aged than the rocks upon which they display themselves.[20]

It was a difficult discussion for autonomists because of their inability to find the language to defend the autonomy of lichens, although Johannes Reinke (1849–1931) came up with alternative terms such as 'consortism' and 'consortium'. He believed the alga and the fungus combined to create 'a common body' that was uniquely autonomous, similar to a human being composed of many types of tissues.[21] Where the dualists saw difference, the autonomists saw unity.

One of Crombie's comments expresses the frustration of integrating lichens into the tree of life: 'Lichens therefore are Lichens and nothing else – neither Fungi nor Algae, or any intermixture of these; but everywhere and constantly preserving their own distinct type, and distinguished by many important characters peculiar to themselves.'[22] Indeed, the lichen symbiosis is the only example in which new morphologies (phenotypes) are created as a result of a partnership between two different species. Perhaps Annie Lorrain Smith said it best in 1921 when she wrote that Schwendenerism, with its emphasis on parasitism, was 'an inadequate conception' for the lichen. She noted that the two partners 'form a healthy unit capable of development and change: a basis for progress along new lines. Permanent characters have been formed which are transmitted just as in other units of organic life.'[23] In other words, lichens act like species, even though they do not fit the definition of species.

The confusing state of the lichen lifestyle inspired a new word to replace Schwendenerism: symbiosis. Bradfield Martin and Ernest Schwab in their review article 'Symbiosis: "Living Together" in Chaos' (2012) write that 'Symbiosis may be the greatest enigma in the history of biological terminology.'[24] They document 130 years of confusion resulting from differing usages of the term that occurred in

When hydrated, the common toadskin lichen (*Lasallia papulosa*), here nestled with a feather moss (top) and acrocarps (bottom), appears green and oozy-looking because the cortex becomes translucent, facilitating metabolic activity of the algal layer.

When mature, the common toadskin lichen, seen here with rock tripe lichen, appears peppered-looking (apothecia) and tattered because growth occurs from the centre of the thallus, causing the edges to tear.

part because lichens were an enigmatic archetype. Schwendener had viewed lichens as a kind of 'good-natured' parasitism, but that undermined the definition of parasitism. In 1877 Albert Bernhard Frank (1839–1900), after studying five species of crustose lichens, advanced the term *symbiotismus* for 'the mere coexistence' of two different species.[25] Frank was careful to state that his term 'does not consider the role which the two individuals play'.[26] In 1879 de Bary introduced a similar word, *symbiose*, as 'the living together of unlike organisms' (*zusammenleben ungleichnamiger organismen*).[27] His term went beyond the scope of Frank's concept, including a panoply of interactions, such as pollination and herbivory as forms of symbiosis. History gives credit to de Bary for coining the term, though he acknowledged Frank's earlier use. The status of the lichen as a dual organism continued.

In 2016 the plot thickened when an image of the forked tube lichen (*Hypogymnia imshaugii*) made the cover of the prestigious journal *Science*, accompanying the lead article 'Basidiomycete Yeasts in the Cortex of Ascomycete Macrolichens'. The lichen was soon termed a 'ménage à trois' in the media. *The Atlantic* carried a nontechnical version of the story with the title 'How a Guy from a Montana Trailer Park Overturned 150 Years of Biology'. The problematical nature of lichens can no doubt be blamed for the somewhat misleading publicity that ensued. The guy from the trailer park, who wasn't literally from a trailer park, was lead author Toby Spribille. Homeschooled and lacking the finances and school records to follow his passion for science at an American university, he worked forestry jobs in Montana until, being fluent in German, he was accepted at the University of Göttingen. On his return he joined the symbiosis lab of John McCutcheon in Montana, where he tackled the problem of how two genetically identical species of epiphytic horsehair lichens could have different phenotypes, that is, observable characteristics.[28] *Bryoria tortuosa* contained high levels of the poisonous pigment vulpinic acid, while *Bryoria fremontii*, called wila, did not and was used as a traditional food by First Nation peoples throughout the Pacific Northwest.[29] Though its major carbohydrate, lichenin, is indigestible, when pitcooked in combination with camassia bulbs and root vegetables it increases the nutritional content of the other ingredients by 23–122 per cent.[30]

A five-year enquiry ensued. Spribille had found a second fungus early on but dismissed it as a 'rogue' contaminant. Several years into the research they went back to this second fungus. Both horsehair lichens had the third partner, but the inedible one had twelve times more of it. Though they found the genetic fingerprint of the second fungus, they could not see the cells. In a cross-section of the lichen they appeared as tiny dots, nearly invisible, hiding among the hyphae of the primary fungus. Using laundry detergent to dissolve the tough outer layer (cortex) of the lichen samples, Spribille eventually found three kinds of *Cyphobasidium* yeasts wrapped in sugars. The lichen was heralded as a threesome by the media, despite Spribille and his colleagues' guarded presentation of their findings.[31] In follow-up research the group documented yeasts in 52 lichen genera, but Spribille stresses that 'they are most definitely not in *all* lichens'.[32] Nevertheless, the yeast connection was hailed as a 'game changer', and the next step was to look for evidence of gene transfer among the partners.[33] The one fungus–one alga and/or cyanobacterium definition of a lichen was eroding as the lichen thallus revealed a host of constituent microorganisms now termed the lichen microbiome.

In 2020 lichenologists David Hawksworth and Martin Grube suggested redefining lichens as complex ecosystems: 'A lichen is a self-sustaining ecosystem formed by the interaction of an exhabitant fungus and an extracellular arrangement of one or more photosynthetic partners and an indeterminate number of other microscopic organisms.'[34] We are told to view the lichen as an 'open system', presided over by a dominant fungus with at least one algal/cyanobacterium subdominant partner and various yeast fungi, living together with other organisms that include bacteria, non-lichen fungi, viruses, protists and archaea. Statistics about the partners are revealing: while 19,400 lichen-forming fungi have been named, so far only 120 lichen-forming algae, which occur in 85 per cent of lichens, have been named. Cyanobacteria occur in 10 per cent of lichen 'species', but only 3 per cent of lichens have both green algae and cyanobacteria.[35]

Lynn Margulis (1938–2011), an evolutionary biologist who provided molecular evidence for the endosymbiotic origin of cellular organelles in 1967, defined symbiogenesis as 'the origin of new organisms, organs, tissues, or behavior traits, as a consequence of long-term symbiosis'. In this connection, she writes of lichens, 'Lichens exemplify the details of complex individuality. The relations between

Brown-eyed sunshine (*Vulpicida canadensis*) owes its yellow colour to lichen acids, Edgewood Blue, Clearwater Valley, British Columbia.

syntropic metabolism and morphogenesis in the emergence of novelty through physical association are made obvious in these colorful creatures.'[36] In other words, the lichen partners create novel appearances (morphogenesis) through the sharing of metabolic compounds (syntropic metabolism). Margulis's comments about the 'emergence of novelty' and the 'complex individuality' of lichens echo today.

Naming complex individuals/ecosystems composed of species from different kingdoms presents problems. Linnaeus grouped lichens as plants in *Lichenes* and gave them binomials, as if they were unique species, such as *Lichen pulmonaria* for lungwort. After the symbiotic nature of lichens was eventually accepted, the International Code of Botanical Nomenclature ruled that lichens be named for the fungal partner for several reasons, but primarily because the fungal partner (mycobiont) comprises 90 per cent of the lichen thallus in most lichens and because fungi that form lichens are usually only found in nature in lichens. So, *Lichen pulmonaria* became *Lobaria pulmonaria*. Trevor Goward, a lichenologist who has tackled what he calls 'the sophistication of lichens' with considerable sophistication himself, discusses the lichen-naming quandary, using as an example a striking lichen whose common name is brown-eyed sunshine and whose fungal name is *Vulpicida canadensis*. It is a lemon-yellow, ruffled lichen, whose 'brown eyes' appear in the form of prominent circular spore-producing structures (apothecia). Goward compares brown-eyed sunshine to a chocolate cake to make the point that although the fungal partner may make up the bulk of the lichen, much like flour in a cake, one does not call a chocolate cake 'flour'. Goward writes that 'Only in common names is the human mind actually permitted unequivocally to touch the lichen thallus.'[37]

The Lichen Life Cycle

The sexual part of a life cycle, where genetic recombination occurs, usually receives the most attention from biologists, but this is problematic in the case of lichens. While the moss life cycle describes perpetuation of single-species individuals, the lichen life cycle must account for replication of a multispecies organism. Like mosses, lichens produce many kinds of minute asexual reproductive units, which disperse both bionts (partners) together, giving rise to clones of the parent lichen. The mycobiont (fungal partner) produces spores sexually, that is, via recombination in distinctive,

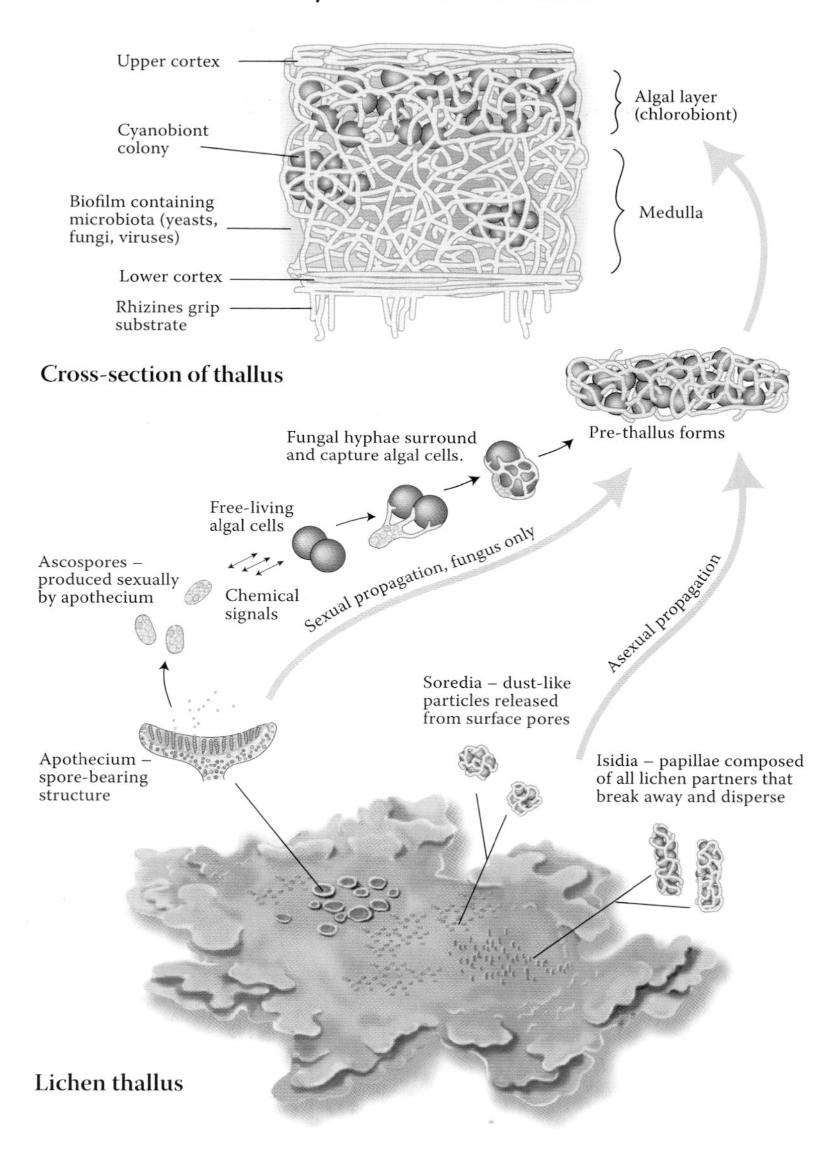

Asexual reproduction predominates in the lichen life cycle; sexual reproduction occurs in each partner separately. Despite millions of years of close contact, horizontal gene transfer between partners has not yet been found.

often decorative structures called apothecia when cup-shaped on the surface, and perithecia when embedded in the thallus, but these spores produce new fungi, not new lichens. The photobionts (algal and cyanobacterial partners) can divide within the thallus, but they are only able to produce gametes (sex cells) outside of the thallus. The sexual life cycle begins when a fungal spore lands on a substrate and encounters its preferred photobiont, an occasion difficult to observe in the field. Cells of both bionts may find themselves in proximity when released into the environment via fecal pellets of lichenivorous slugs and mites or via lichen fragmentation. Creation of a new lichen thallus by partners reassociating in the field is termed 'reassembly', or 'resynthesis'. Researchers follow a progression of stages from precontact to physical contact to disordered prethallus to mature layered/branched thallus. They describe resynthesis as a 'knitwork' mediated by 'chemical signalling' or 'molecular crosstalk' between the bionts.[38]

For many years the question of how often lichens reassemble *de novo*, that is, not from asexual fragments but from the bionts 'dating' each other anew, has been a black box. However, a 2012 study of 62 populations of lungwort (*Lobaria pulmonaria*) revealed that while 70 per cent of both bionts were clones, recombination occurred in almost 8 per cent of the mycobionts. The photobiont did not show statistically significant genetic recombination, but there was evidence of somatic mutation. Lichenologists Martin Grube and Toby Spribille write that this represents 'a highly successful balance of safe recruitment of symbiotic clones and endless possibilities for fungal recombination and symbiont reshuffling'.[39] In 2014 William Sanders depicted all aspects of the life cycle of the leaf-dot lichen *Calopadia puiggarii*, a widespread leaf epiphyte found in warm climates. The painstaking work, which began in 2008, involved tying plastic cover slips on sabal palmetto leaves and collecting them at various intervals for study under the microscope. The sexual and asexual parts of this life cycle involve many moving parts due to several kinds of spores and numerous propagules.[40]

Biologists describe the lichen thallus as an 'edifice' whose construction represents a 'mysterious transformation'.[41] But Goward argues that lichens do not grow as such; rather, they are built.[42] Recognition and chemical signalling begin pre-contact. Gene activation occurs in both partners, with a subsequent release of compounds into the space between them. Key ancient hormones (auxins, cytokinins, ethylene and salicylic acid), whose occurrence in mosses and lichens pre-dates their presence in higher plants, or lectins, glycoproteins that occur in all forms of life, may be at work here. In one hypothetical scenario lectins diffuse into the space between the prospective partners and encounter chemical messengers (ligands), which leads to a cascade of reactions resulting in initiation of physical contact or death of an incompatible photobiont cell.[43] In successful interactions the fungal hyphae grow towards the accepted photobiont cell and proceed to 'enwrap' it, establishing wall-to-wall connections called 'junctions'. The photobiont increases in size when enwrapped, a sign of well-being noted by Schwendener and Thwaites. The junctions allow passage of metabolites from the alga that initiate the lichen-building behaviour of the mycobiont hyphae. Although the mycobiont has often been given the greatest credit for the construction material, the two-by-fours making up the scaffold, the photobiont cells, though far fewer in number, are considered full partners of the edifice in that they supply the liquid blueprint for the building process. Goward uses the word 'modular' to describe some aspects of lichen construction (see earlier for modular growth of mosses).

Ursula Goodenough of Washington University, working with a team of other researchers, studied the development of thallus structure in several lichens with electron microscopy. The synthesis of the thallus begins in an area of loosely woven hyphae called the medulla. Some hyphae bifurcate, extending linearly, while other hyphae diverge at right angles to the mother hyphae. These have specialist functions. Some proceed to the surface to form the protective outer layer called the cortex. Cortical hyphae secrete polysaccharides that act as glue, forming a 'conglutinated' topmost layer. Most lichens have this protective upper cortex, but patterning details vary. Other hyphae form what Goodenough calls the 'green modules' of the algal layer, which is directly under the upper cortex. Here more junctions, so essential to biont communication, are created, as photobiont cells divide and hyphae form sub-branches to enwrap these clusters of cells. Some form soredia, powdery asexual propagating units on the surface of the thallus, while others, as in the ruffle lichens, form vents to the surface for gas exchange. Eventually some hyphae become acellular, serving structural functions as struts and honeycombs. The knitwork must be cohesive, in order to withstand mechanical stress, yet loose enough to allow for hygroscopic flexibility during moisture absorption. Algal cells have shown a capacity for cell wall remodelling during

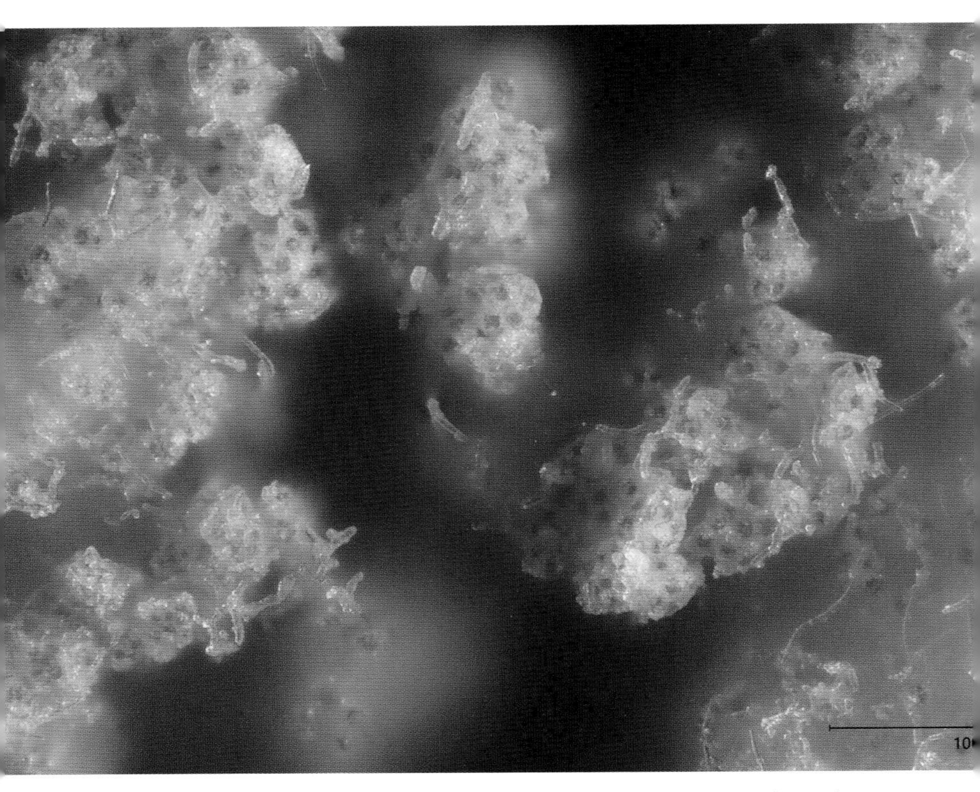

Martin Grube calls his image of the soredia of the lichen *Lepraria* 'symbiotic dust', emphasizing the vast potential of asexual dispersal in lichens.

these cycles. 'The formation of tight yet swellable fungal layers' that can respond physically to repeated cycles of desiccation and rehydration is considered a key innovation in the evolution of the lichen.[44]

Lichens at all stages are covered in bacterial biofilms that physically participate in lichen construction and attach lichens to bark and other substrates.[45] Many develop from bacterial rain; others, such as cyanobacteria, are captured from mosses, according to studies of lungwort (*Lobaria pulmonaria*).[46] Bacterial biofilms themselves are multispecies constructs with different pieces fitting together, discretely, like pieces of a puzzle.[47] Often the bacterial component is lichen-specific. Goodenough et al. term the biofilms in the medulla formed by all lichen microbiota a 'fog'. They describe it as a viscous fluid in a glass-like phase that acts as an antifreeze, offering protection during desiccation.

Goward refers to the lichen thallus as 'a solar panel designed for optimum display of the lichen photocells within'. Indeed the mycobiont is able to adjust the position of the algal cells over short distances to maximize photosynthesis.[48] Compartmentalization of the competing activities that occur side by side in the interior of the lichen, such as gas exchange for photosynthesis versus transfer of nutrients in water, are made possible in part by hydrophobins. These are unique proteins produced by all filamentous fungi, which occur in most lichens. The molecules are amphiphilic, one side hydrophilic and the other side hydrophobic. Hydrophobic sides of the molecules surround the green modules where gas exchange occurs, while hydrophilic sides face the upper cortex, where quick absorption of water is important. Goward also reminds us that crucial parts of the thallus function as 'carbon boosters, carbon quenchers, carbon conductors, carbon sinks, and carbon depots'.[49] All fungi are heterotrophs, dependent on other organisms for food, while green algae and cyanobacteria are autotrophs, able to feed themselves through photosynthesis. According to the nutritional model of lichen symbiosis, the mycobiont receives carbohydrates in the form of sugar alcohols from a chlorobiont (green algae) and glucose from a cyanobiont (cyanobacteria). In turn, the mycobiont provides shelter, minerals and a host of carbon compounds sometimes called 'secondary compounds' because they are not required for basic metabolism. These lichen acids, often found in crystallized form on the surface of or inside the thallus, may have roles in lichen construction and maintenance, but certainly protect the thallus in various ways and enhance fitness. Their copious production, sometimes up to 30 per cent of their body weight, is seen as an extraordinary outcome of lichen symbiosis. Recently Spribille and colleagues have suggested that 'the reciprocity that stabilizes' the lichen is less about nutrition and more about enhancing ecological fitness. They note, 'Lichen fungal symbionts count among the only filamentous fungi that expose most of their mycelium to an aerial environment.'[50]

The 'open system' of the lichen permits considerable scope for improvisation in the field to circumvent the bottleneck of resynthesis and enhance fitness. Sometimes the mycobiont will accept a substitute partner, most likely in the same genus, and make do until the desired partner is nearby. Partner switching may be either transitory or permanent.[51] Many non-lichen-forming fungi live on lichens. Some will employ a strategy of kleptosymbiosis (photobiont capture) in which their fungal hyphae retrieve an algal cell from the thallus below, and then create their own self-sustaining thallus. Some thalli contain more than one genotype of mycobiont and photobiont, which may represent intermediate stages in symbiont resynthesis or evolution in progress.[52] In their survey of lichen 'symbiont management', Grube and Spribille state that 'to adapt to local climatic conditions . . . lichens can adjust the ratio of algae as producers versus fungi as consumers in their thalli' and, further, that they 'can select locally optimized strains or species for thallus formation'.[53] A similar study of lichens 'at the edge of life' in three valleys in Antarctica found a general trend towards low selectivity in the mycobiont–photobiont relationship, but specialist relationships survived even in that harsh environment.[54] Thalli of the lichen *Ramalina farinacea* consistently show the presence of two types of *Trebouxia* green algae that perform best at differing levels of temperature and irradiation, suggesting that the lichen will have an advantage in a greater range of ecological settings.[55]

Lichen Novelty

Estimates of the number of lichens, which show astonishing diversity in form, range from 14,000 to 20,000. Acharius established three groupings – crustose, foliose (leafy) and fruticose (shrubby) – that are still used today to organize this diversity, which Goward has subdivided in more detail. His short list (a longer one exists) highlights the novelty of lichens: gels, leathers, fires, dusts, crusts, scales, navels, leafs, mantles, suedes, stickpins, pixies, scrubs, hairs and shags. He reminds us that this morphological diversity is a response to place, and that we must learn 'to read the lichen thallus as a surrogate for [it]'.[56]

Seventy-five per cent of all lichens are crustose species, which grow tightly appressed to their many possible substrates, natural and man-made. Often mistaken for odd blotches on a rock, they can take the form of films, dots and dusts. Some are puffy (bullate), some are effigurate (marginal lobes prolonged and radially arranged) and others are endophloedic (living on the surface of plants). Some even infiltrate rock, becoming rock-hard themselves, needing to be scraped off with a chisel for study. They build soil, slowly, by degrading stone with acids. As the thallus expands and contracts during wetting and drying, mineral grains are detached infinitesimally. The millions of years of lichen weathering of the Earth prompted mycologist Merlin Sheldrake to write, 'A portion of the minerals in your body is likely to have passed through a lichen at some point.'[57] Sometimes thalli of different crustose species fuse, forming mosaics.[58] The map lichen (*Rhizocarpon geographicum*) is an iconic member of this group; it forms close communities, with blackened fungal hyphae marking the boundaries. It is commonly used in lichenometry (dating geological structures based on lichen size and growth rate).

About eight hundred or so species of crustose lichens live as epiphytes on the leaves of tropical and subtropical trees. Unlike those

Orange lichen brings life to the crevices of limestone boulders in western Virginia.

that live on rocks, they must grow quickly to keep pace with their ephemeral substrates. In general, crustose lichens have just two layers, the cortex and the medulla. Despite this minimalism, they adapt their 'edifice' to extreme conditions. Species that live in the South African desert have a cortex six times thicker than that of the same species found in Europe. The jellyskin lichens, in which cyanobacteria are the

Crustose lichens form a mosaic on a stone wall, Highland County, Virginia.

dominant partner, have no layered structures, just blue-green *Nostoc* cells mingled among fungal hyphae in a jelly-like mass that becomes rigid when desiccated.

For foliose and fruticose lichens, often called macrolichens, place means community as well as environment. The foliose lichen *Lobaria pulmonaria* (lungmoss, lungwort), the most well-studied lichen in the world, was so famous for the epiphytic community it dominated on the trunks and major boughs of trees in old-growth, undisturbed forest in many parts of the world that the community was given a Latin name, *Lobarion pulmonariae*, now more simply referred to as the Lobarion.[59] These ancient communities arose in benign settings where environmental variables such as temperature, humidity, light exposure, air quality, speed and frequency of wet–dry cycles and bark pH supported luxuriant growth. Equally important were community members such as its four relative species, other lichens, mosses, microfungi, algae, cyanobacteria and many microinvertebrates. Lungwort prefers trees over a hundred years old, when bark pH is less acidic. Its light preferences mean that it needs open woodland

The antlered jellyskin lichen (*Scytinium palmatum*), which stabilizes disturbed soil and fixes nitrogen, changes colour and expands when hydrated.

Blue lungwort (*Lobaria scrobiculata*) grows on a birch tree in the company of various mosses in Tomies Wood, Killarney National Park, Ireland.

rather than closed-canopy situations. Its vigorous growth and ribbed, reticulate surface suggest to Goward that it must have a transport system for moving nutrients (vasculature). In *Lichens* (2000) Oliver Gilbert described the (former) abundance of a Lobarion near a loch in an area of temperate rainforest in Scotland. Thalli the size

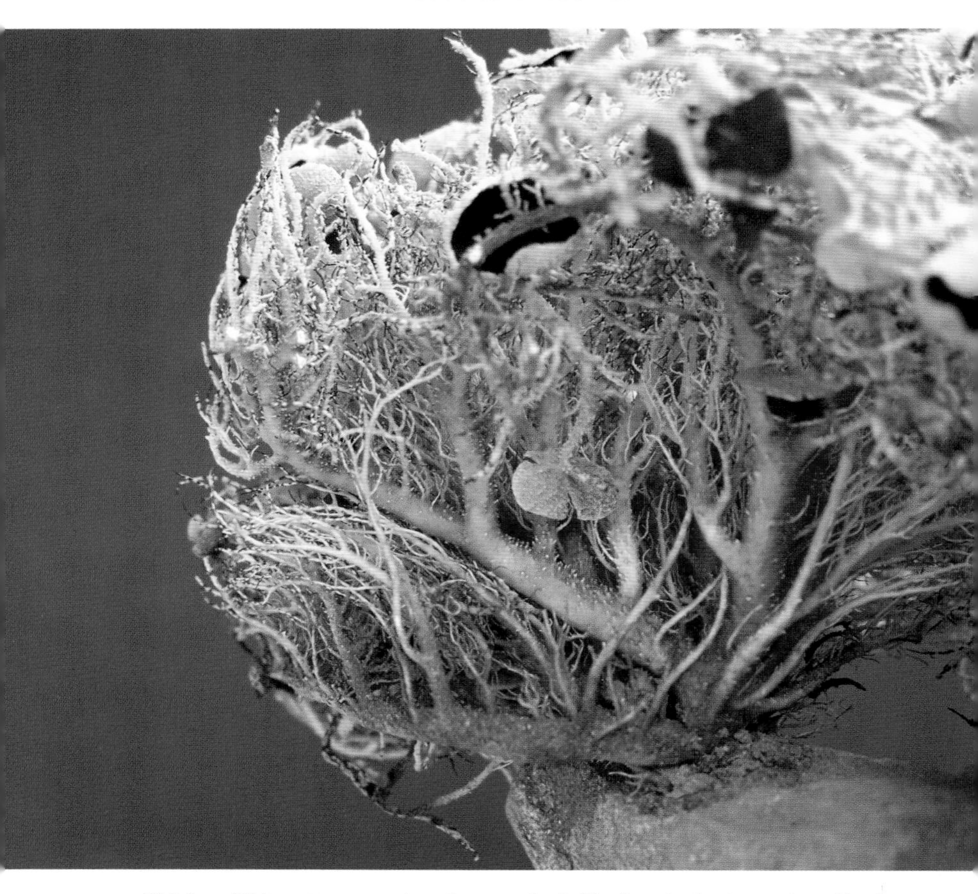

This beard lichen, *Usnea antarctica*, takes on a shrub-like form in the Antarctic, unlike the pendant species in temperate climates; it adheres to rocks with a holdfast.

of dinner plates hung from trees, falling off in heaps at their base. The loose attachment of lungworts by a single holdfast increased gas exchange, supporting faster growth.[60] These communities are considerably diminished today from industrial logging, air pollution and loss of habitat. Transplant experiments to move individuals from felled trees or endangered habitats are under way in many countries. A recent transplant effort occurred in 2020 in the Lake District, after a veteran oak, which held the largest Lobarion in England, blew down in a storm.[61]

Fruticose lichens begin as prostrate thalli but transition to semi-erect three-dimensional forms reminiscent of higher plants. Some, such as reindeer lichens (*Cladonia* sp.) and foam lichens (*Stereocaulon* sp.), are mat-forming genera that attach to the ground and account for 97 per cent of the ground cover and 20 per cent of the total ecosystem biomass in the open woodlands of northern Canada.[62] Other fruticose species take the form of pendant 'beards' and lacy curtains, such as *Usnea* and *Ramalina*, which favour humid areas but find refuge in 'sky islands', peaks 1,828 metres above the surrounding deserts of southwestern Arizona and northern Mexico.[63]

'Thinking about whole lichens isn't always easy,' writes Goward, although it is his forte. By 2012, he had moved beyond his description of lichens in the 1990s as fungi that have learned to farm, to recognition of their status as unique emergent organismal ecosystems.[64] Merlin Sheldrake writes, 'Lichens are stabilized networks of relationships; they never stop lichenizing; they are verbs as well as nouns.'[65] 'If we define an organism as a self,' writes Ursula Goodenough, 'then a lichen meets the criteria: it self-organizes, it self-maintains; it self-protects, and it self-reproduces to form more selves of the same kind.'[66] Lichens have colonized the terrestrial world, she says, because 'they bring their self-sustaining ecosystems with them and because they aren't in any hurry.'[67] In any event, as lichenologist Rosmarie Honegger notes, 'the dual hypothesis of lichens opened the minds of scientists far beyond lichenology.'[68] This dynamic continues. One remembers Reverend Crombie's passionate assertion that 'Lichens therefore are Lichens and nothing else.' We might amend Crombie's statement to 'lichens are lichens – and *something* else.'

four
Cosmopolitan Extremophiles

> Fyfield Down . . . [near] Marlborough is moss and lichen heaven – the mysterious sarsen stones are home to lichens and rare mosses . . .
>
> TWEET FROM @NATUREBUREAU, 31 January 2021

Mosses and lichens call many places on Earth heaven, even those that are degraded, unnatural and impermanent. Protected habitats like the ancient sarsen stones in the UK are wonderful places to observe their centuries-old growth, while bare ground and tarmac in petrol stations and malls, where they bring life to soil littered with bottle caps and cigarette butts, showcase their work as colonizers. Nature's graffiti, they are nondiscriminatory in their artwork. Their cosmopolitan distribution is impressive: lichens cover approximately 8 per cent of the world's terrestrial surface, while mosses are responsible for above-ground net primary productivity (NPP) of 20 per cent in boreal forests and 50 per cent in wetlands, on average.[1] Although their ubiquity implies an ability to generalize, they are the ultimate specialists. Both are termed extremophiles because they survive in hostile habitats and extreme climates. Much of their success lies in how they manage their inability to store water. Mosses and lichens spend much of their time as still life, waiting to reanimate after long periods of desiccation.

Lichens enliven boulders on the shores of Loch Scavaig, with a view of the Isle of Skye in the distance.

'Between a Rock and a Dry Place': The Art of Desiccation Tolerance

In taking on the supreme challenge of living without water, either intermittently or for the long term, mosses and lichens join a group with an unwieldy name – the poikilohydric anhydrobiotes. *Poíkilo-* is from the Greek *poecílos* for a many-coloured, spotted, embroidered, variable or irregular condition, while *-hydric* refers to water. The name suggests their ability to adapt to a varied water supply. When fully desiccated these organisms enter a state of suspended animation where metabolic activity is undetectable. It is a fundamentally different strategy from that adopted by higher plants and most animals. Instead of defying external environmental conditions, they coequilibrate with their environment. Mosses and lichens endure a roller coaster of wetting and drying cycles that are often sudden and extreme. Bryologist Brent Mishler writes that mosses have 'the same water relationships as a piece of paper sitting on a rock – if it rains it gets wet fast, if it stops raining it gets dry fast'.[2]

Mosses and lichens face serious water fluctuations daily, even hourly.[3] When higher plants are faced with sustained dehydration, a series of devastating biochemical changes occurs at the cellular level. The water shell around macromolecules disappears, and membranes denature. Death follows. Mosses and lichens, instead, have developed a series of specialized metabolic responses, a form of triage known as desiccation tolerance (DT), which enables them to wait out the period of dehydration. Evolutionary biologists studying how aquatic life forms first adapted to terrestrial existence believe that the first plants to abandon life aquatic for life terrestrial must have had mechanisms for DT in place. When mosses and angiosperms diverged from their common ancestor, the former retained those metabolic pathways, and the latter abandoned them, except perhaps in seeds, though the renowned DT of seeds could be a re-evolved phenomenon. This divergence resulted in different growth outcomes. Organisms with DT have a restricted metabolism, with long periods

Dagoda domes, which store sacred relics, provide substrate for mosses, Kiri Vihara temple ruins in Polonnaruwa, Sri Lanka.

of enforced fasting due to metabolic inactivity: they grow slowly and remain small. Having sacrificed DT, higher plants benefited from their more active metabolism to increase carbon gain – and size. Larger size, however, restricts their choice of habitat.[4] DT is a primitive strategy that remains important in maintaining life on land.[5] Evidence shows that desiccation tolerance is a dynamic response: mosses and lichens calibrate their biochemistry from season to season in response to their environment. Further, there is evidence that lichenization enhances the DT of each partner.[6]

Desiccation tolerance involves two phases: the ability during extreme conditions to swiftly halt metabolism (for example, photosynthesis and respiration) while preserving basic cellular components; and the ability to quickly resume metabolism and restart growth when conditions improve. Upon being air-dried at the cellular level, they enter a latent 'glassy' state, also called suspended animation, anhydrobiosis, cryptobiosis or anabiosis, in which molecular motion slows to a near standstill. New techniques are allowing researchers to study enzyme activity in the glassy state. Recent research on desiccation tolerance in mosses and lichens is focused on the interplay between constitutive and inducible cellular processes during dehydration and subsequent rehydration. Constitutive mechanisms are those that remain poised in the glassy state in the absence of cellular water, while inducible ones are those that need to be activated *de novo* upon hydration. Researchers believe that in lichens the constitutive pathway is more important because it is imperative for them to get a fast start on photosynthesis and subsequent carbon gain.[7] The fungal partner, which makes up 90 per cent of lichen thalli, is dependent upon the sugars provided by the algal partner for respiration and growth.[8] Desiccation tolerance also provides protection from extremes of temperature, intense sunlight and ultraviolet radiation.[9]

Mosses and Lichens Colonize The World

Mosses and lichens colonize (Latin *colere*, to inhabit) bare earth (terricolous), rocks (saxicolous), bark (corticolous) and leaves (foliicolous) in tropical and subtropical habitats, the occasional animal and every conceivable man-made surface. Bryologists and lichenologists visit unusual places in search of specialist communities, such as the toxic shadows under pylon lines: water dripping off metal creates acid conditions in which only certain mosses and lichens grow.[10] Images of a lichenized school bus found in southwestern Australia show the glass headlights, plastic tail-light cover and metal body colonized by a species of the lichen *Xanthoparmelia*.[11] Lichens tolerate many substances toxic to other organisms (such as lead, zinc, arsenic, copper and uranium minerals) and incorporate oxalates of metals (such as copper, manganese and magnesium) into their thalli without harm.[12] Natural substrates offer more nuanced microhabitats. Lichens distinguish among many different aspects of bark: acid bark, mesic bark, flushed bark (beech), rough bark, smooth bark, ancient dry bark, twigs and lignin. Bryologist Jerry Jenkins mentions numerous ecological settings that specialist mosses occupy: dripping rock faces, seepage cracks, wet ledges, shaded boulders versus dirty boulders, young trees with tight bark and small crowns, older trees with rough bark, living trees, logs and stumps.[13] Jenkins takes online observers to the 'stony impervious soils' of railway lines with a discussion of how 'habitats select their mosses, and how mosses manage their habitats'.[14]

Lichens that colonize glass are termed vitricolous. In 1922 Ethel Mellor of the University of Reading studied the lichen flora of church windows in France. She found 24 species, mostly crustose lichens.

Wistman's Wood in Dartmoor, Devon, home to many epiphytic mosses and lichens, was declared a Site of Special Scientific Interest in 1964.

Mosses and lichens colonize vast areas of Iceland's lava fields; seen here are the woolly fringe-moss *Racomitrium lanuginosum* and the lichen *Cladonia uncialis*, Vík í Mýrdal, Iceland.

When the glass becomes too pitted for crustose species, leafy lichens take over. Mellor describes how CO_2 produced in lichen respiration added to CO_2 dissolved in water trapped between glass and lichen thallus produces chemical changes in both glass and leading. Further, Mellor demonstrated that these lichens showed sensitivity to the different colours of stained glass, following the pattern of certain colours while ignoring others.[15]

The amateur daunted by moss and lichen identification can learn about their ecology by observing where they grow. With trees it is best to start at ground level (the tree bole) and look up in stages. As mosses and lichens often grow on the same tree, the issue of competition arises. Gaining information about competition requires sampling over the entire length of a particular tree. The type of tree determines key variables such as bark pH, aspect and texture, while height above ground level determines varying effects of wind, humidity and light intensity. Further, competition mediates succession as mosses and lichens respond to the continuous change in available niches. Mosses are faster-growing and taller, but lichens can tolerate less humidity and more intense light and can use secondary compounds like vulpinic acid to defend space. In similar niches, offensive foliose lichens compete with defensive crustose lichens. In areas where two species co-occur, creating a highly antagonistic environment, researchers found 'truce margins'.[16]

Although some say that lichens prefer to grow on steady surfaces, *vagrant* lichens of deserts, prairie, chaparral and paramo are designed for wandering. Vagrant lichens are not to be confused with *erratic* lichens, which prefer to grow attached but can adapt to upheaval. True vagrant species such as *Aspicilia fruticulosa* and *Xanthoparmelia convoluta* have thallus margins adapted for rolling up into balls. Vagrant lichens rely on winds to carry thallus fragments around the world. Mosses and lichens can lodge on moving targets as well. In New Guinea the backs of large Papuan moss-forest weevils carry a well-stocked flora, including various species of lichens.[17] Some insects use saliva to glue fragments of moss and lichen to their bodies. Tiny lacewing larvae cover themselves with lichen and moss fragments for camouflage.[18]

Structural adaptations for absorbing water in any form – rain, snow, mist, fog or dew – mediate in low-rain situations. The cushion form's dense tufts and clumps reduce evaporation by creating a rough boundary layer. The dramatically elongated leaf tips of the gametophytes, also called awns or hairpoints, reflect the sun and reduce energy absorption, temperature and water loss.[19] The fluffy awns roughen the boundary layer where the surface of the moss meets the air, reducing evaporation by 20–35 per cent, thereby trapping stagnant, ideally moist air. Also, their extreme length, in proportion to the height of a

Bright red apothecia of British soldier lichens (*Cladonia cristatella*) stand out in this moss-and-lichen community on the boulder fields of Back Creek Mountain, Virginia.

moss gametophyte, increases the distance from the actual surface of the leaf to the air, which decreases the diffusion gradient, slowing the process of evaporation. Lichens have many ways of increasing surface area for water absorption. Foliose lichens increase surface area with lobes, frills and ridges, while fruticose lichens can be pendant, such as the witch's-hair lichen (*Alectoria sarmentosa*) and old man's beard lichen (*Usnea longissima*), which are several feet long. Lichens containing blue-green algae (cyanolichens) are especially efficient sponges, absorbing as much as 2,100 per cent of their dry weight in water.[20] Both mosses and lichens can drown from water-logging, however. Prolonged excess water deprives the gametophyte or thallus of oxygen (anoxia). There are only a few aquatic or partially aquatic mosses, and, of the 250 freshwater and 700 marine tidal lichens, most are amphibious. Many of these lichens belong to the family Verrucariaceae (from the Latin *verruca*, wart). Research conducted at the Natural History Museum in London seeks the genetic basis for survival in conditions where carbon dioxide and oxygen are in limited supply.[21]

Biocrust

Mosses and lichens play a dominant role in biocrusts, often called the 'living skin' of the Earth. Enmeshed within gelatinous bundles of filamentous cyanobacteria, green algae and liverworts, they comprise a biologically active interface between the atmosphere and the soil, primarily in arid, cold and low-productivity environments. Previously disregarded as scum – soil covered with biocrust is noticeably darker due to UV-protective pigments of cyanobacteria – their global relevance in carbon and nitrogen cycles is now recognized. As absorptive cover, biocrusts shelter microarthropods, manage water relations, prevent erosion and influence reflectance of light (albedo), thereby affecting soil temperature. Biocrusts are surprisingly effective in trapping aeolian dusts, the nutrient-rich particles driven round the world by winds. As the value of biocrusts has been noticed, so has their decline. They form slowly and when deformed by compression (trampling, footprints, grazing) are slow to heal. Restoration methods such as replanting damaged areas with material from biocrust nurseries are being evaluated.

Moss and Lichen in the Desert

Approximately equal numbers of mosses and lichens occur in deserts, where new species continue to be discovered.[22] While they may flourish in fog deserts such as the Atacama of Chile, the Mojave of North America and the Namib of South Africa, the moss *Syntrichia caninervis* forms extensive carpets on the inland Gürbantunggüt Desert of northwestern China, the desert furthest from any sea.[23] When hydrated, the colour is a reddish-greenish-brown, but when dry it goes into disguise mode with shrivelled leaves spiralled tightly around the stems, darkened to near black.[24] Volkmar Wirth has photographed many of the Namib's 250 lichen species.[25] A new species described in 2007 from the Namib finds habitat *inside* small limestone pebbles. Lichens infiltrate rocks in many ways: epilithic lichens live on rocks; chasmolithic lichens grow between mineral grains of the surface; cryptoendolithic lichens find pre-existing cavities inside rocks; and euendolithic lichens form new cavities inside rocks by dissolving the substrate. *Buellia peregrina*, newly described from the Namib in 2007 and named for its novel pigment, joins the small group of euendolithic lichens.[26]

When bryologist Lloyd R. Stark left the mossy lands of Vermont for the bio-crusted landscape of the Mojave desert, he was surprised to find approximately 125 species of mosses tucked away in the shade of north-facing boulders and creosote bushes, where they endure nine months without rain, on average, and where too much rain at the wrong time can be more dangerous than too little. A sudden thunderstorm in summer can raise temperatures to between 32 and 38 degrees Celsius, which, together with the sudden waterlogging, can be lethal. Stark found that desert mosses with separate male and female

plants have one male for every one hundred females, the most biased sex ratio known for any plant. In fact, all-female populations were common. Moss sperm are produced in packets that contain hundreds of sperm, which use more resources than the female's one or two egg cells. Female plants were found to respond to high stress by spontaneously aborting offspring. Researchers are also investigating whether female plants might be more desiccation tolerant than males.[27]

In 2014 Kirsten Fisher, a biologist at California State University in Los Angeles, was studying the sex life of *Syntrichia caninervis* in the Mojave desert. It is one of the most common mosses there and has a lifespan of about one hundred years. The research was slow because, as Fisher said in an interview, 'They're never having sex. For a moss, it's maybe once every 30 years.' While doing transect sampling, Fisher turned over one of the many quartz pebbles in the area. Underneath there was a bit of moist green moss. This discovery

Right: Lichens form extensive colonies in the desert near Wlotzkasbaken, Dorob National Park, Namibia.

Left: The Bay of Fires, Tasmania, was named for the beach fires of Tasmanian Aboriginal peoples, which English navigator Tobias Furneaux observed in 1773 from aboard ship, rather than the lichens.

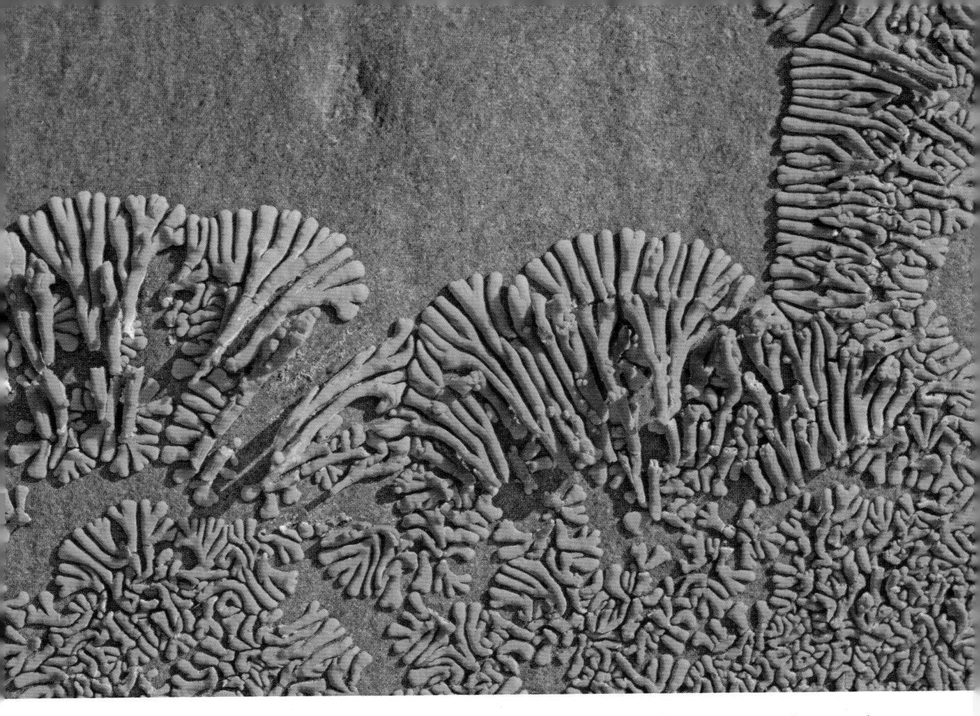

Caloplaca elegantissima, a firedot lichen that occurs worldwide, is native to the Namib. A new Californian species, *C. obamae*, was named after Barack Obama in 2009.

led to the description of a new microhabitat termed quartz oases.[28] Light transmission of about 4 per cent through the milky quartz is just enough to allow photosynthesis while preventing damage from intense solar radiation. Another advantage is the greater humidity found under the protective parasol. Other desert mosses, like the screw moss *Tortula inermis*, were found sheltering with *Syntrichia caninervis*. The description of a new microhabitat is exciting for ecologists who study evolutionary refugia and ecological refuges in the context of conservation.[29]

Chinese researchers have made marvellous discoveries about the moss awn and water collection in *Syntrichia caninervis*. While the awn is deceptively insignificant, just 0.5 to 2.0 millimetres long with a slight axial twist, years of data collecting predicted its importance.[30] Although awns make up only 4.8 per cent of shoot weight, their presence increases water content by 24.9 per cent and delays dehydration by 20 per cent. Plants without awns display stunted growth.[31] But it was collaboration with fluid dynamics expert and engineer Tadd Truscott of the Splash Lab at Utah State University that revealed how the architectural design of the awn functions in water collection. *Syntrichia caninervis* has unique micro-design features whose function is to trap water molecules at different scales: grooves on the awns at nano- and micro-scale nucleate water droplets from a humid atmosphere, while barbs clustered in certain areas collect fog droplets, moving them along at 'impressive speeds'. Further, the flexible awns and leaves absorb raindrop energy, which reduces splashing and water loss. This 'anti-splash raindrop absorption mechanism' helps mosses at the edge of a clump, which are the most vulnerable but also the most capable of growth. Engineers are now fabricating miniature *Syntrichia caninervis* bionic structures that reduce splashing and enhance water spreading.[32]

Mosses and the frosted grain-spored lichen, which infiltrates rock surfaces leaving only the apothecia (black dots) visible, share a boulder, Ridgefield, Connecticut.

Mosses and Lichens at the Poles

Mosses and lichens are the dominant photosynthetic cover in both the Arctic and Antarctic, partly because they can colonize these areas in the absence of higher plants and partly because the conditions suit their toolbox. Sometimes the term 'polar desert' is applied to these regions, which have constantly low temperatures below −10 degrees Celsius (and physiological drought due to ice). While mosses in deserts insulate the soil against high temperatures, at the poles

This typical moss turf from east Antarctica includes the endemic *Schistidium antarctici* (olive green in colour), which is the dominant moss species in the area. The larger, pale green-brown patches are *Bryum pseudotriquetrum*, and the red moss in the background is *Ceratodon purpureus*. Lichens on the rocks include *Usnea*, *Umbilicaria* and *Pseudephebe*. The moss bed is fed by melt water from the snow bank.

they insulate against thawing of the ice.[33] Lichens can photosynthesize down to –15 degrees Celsius. About 350 species of lichens and 115 species of mosses, but only 2 species of higher plants, live in Antarctica, described by ecologists as the driest, coldest and windiest continent on Earth.[34] Researchers are studying mosses and lichens as key players in the carbon and nitrogen cycles in high-altitude ecosystems threatened by climate change.[35]

The poles offer a range of habitats for moss and lichen specialists. Nunataks (Inuit for 'lonely mountain') stud the ice fields of inland areas and are considered refuges of Antarctic diversity.[36] These ice-free, exposed rocky outcroppings may be isolated, serving as landmarks, or clustered in groups. As many as ten endemic species of lichens have been identified on the Utsteinen Nunatak of the Sør Rondane Mountains in Antarctica, where a Belgian Antarctic research station monitors grids established in 2007 for collecting data on climate change effects on lichen populations.[37] The cosmopolitan moss genera *Grimmia apocarpa* and silver moss (*Bryum argenteum*) are also found on nunataks, prompting biogeographers to study how they found, settled and diversified in these habitats of 'almost unbelievable severity'.[38]

The Antarctic's marine coastal habitats are seen as test cases of 'symbiosis at its limits'.[39] One study involves the borderline lichen *Mastodia tessellata* (Verrucariaceae), which has a bipolar distribution. It presents an unusual situation in that a free-living algal partner (*Prasiola*) coexists with a 'lichenized' form (*Mastodia*), a situation that might be described as symbiosis in medias res. Several borderline lichens in marine areas seem to involve fungi, just now becoming known, that may represent 'an old lineage reminiscent of early forms of lichen symbioses that indicate the transition from an aquatic habitat to a terrestrial environment'.[40] Speculation continues about whether lichens slid out of water or formed on land, or both. The fungal hyphae of some of these borderline lichens are thought to have rock-boring abilities.[41] Another study in Antarctica compares the photosynthetic capacities of crustose, foliose and fruticose lichens. Fruticose lichens had the greatest carbon gain, indicating that although the fungal partner cannot photosynthesize, it orchestrates greater photosynthetic capacity as the architect of the convoluted fruticose growth form.[42]

As described earlier, some lichens live an unattached lifestyle, becoming vagrant or erratic depending on circumstances. Some

Mosses and lichens add colour to a chinstrap penguin's backdrop in Antarctica.

mosses can do the same, their tufts becoming spherical when dislodged. Moss balls found in extreme environments stroll, sometimes even gallop, in semi-choreographed herds across the bright whiteness of glaciers.[43] Their playful presence continues to evoke delight, fondness and professional interest in glaciologists.[44] Explorers in remote areas worldwide have applied a host of descriptive names. In 1951 Icelandic researcher Jón Eyþórsson bestowed their most popular name: '*jökla-mýs*', or glacier mice.[45] The first recorded sighting occurred in 1874 during observation of the transit of Venus when USS *Swatara* visited Desolation Island, one of the South Shetland Islands near Antarctica.[46] Theories abound about these 'mice on ice', described by one traveller as 'as close as anything in the plant kingdom comes to being a mammal'.[47] There is some consensus about a few characteristics: they form around small pebbles or fine aeolian dust; they can contain several species of moss; they can roll spontaneously without being driven by wind (recorded by accelerometers); and they contain thriving invertebrate communities.

That several generations of springtails can exist together in moss balls indicates that they breed inside the balls, allowing invertebrate fauna to colonize glacier surfaces where they find refugia. This has alerted glaciologists to the 'surprisingly complex ecology' of the glacier surface.[48]

Lilliputian Worlds

Mosses and lichens live in close community with fellow poikilohydric anhydrobiotes, a dazzling array of microinvertebrates that includes mites, springtails and myriapods in Arthropoda; tardigrades or water bears in Tardigradia; and nematodes in Nematoda. All are extremophiles.[49] Arthropods represent an ancient lineage that has coevolved with mosses for millennia.[50] Welsh palaeobotanist Dianne Edwards describes fossil evidence dating to the Devonian/Silurian time period (417 mya) that demonstrates interaction between myriapods and host plants.[51] Equally cosmopolitan, these minibeasts go where mosses and lichens go, thriving in the alpine and polar regions where higher plants cannot survive. Though exceedingly abundant in all sorts of surprising places, their microscopic size hinders appreciation of their biology.[52]

The aquatic genera – tardigrades, rotifers and nematodes – live in the water films that cover mosses and lichens in moist conditions. Terrestrial genera such as armoured moss mites dwell in damp, dark places favoured by mosses but also follow epiphytic lichens high into tree canopies.[53] In a study of arboreal mites living on lichens in sugar maple canopies of the Adirondacks in the United States, researchers found that mites respond to minute details of a lichen's shape, texture and colour. For example, the ruffle of a foliose lichen thallus will afford more moisture than a crustose one, and despite resilient mechanisms, some water is preferred to no water. Mite populations were seven times greater on lichens than on bare tree bark.[54] Mosses and lichens growing together often share mite populations. Mites nibble on both, but in return for shelter disperse asexual fragments of their hosts. Various 'bugs, aphids, and mites suck out the contents

Geoscientists describe lichens as 'extraordinary ice nucleators' that travel the world airborne, markedly influencing atmospheric processes (cup lichen, *Cladonia* sp.).

of moss cells', but for the most part, algae, bacteria and archaea (ancient single-celled organisms that lack a cell nucleus, formerly called prokaryotes) form the bottom of the food pyramid in these busy communities.[55]

The tardigrades – literally 'slow walkers' – also known as water bears or moss piglets, are found in great numbers on both mosses and lichens. One researcher reported 2,287,000 individuals per square metre of silver moss (*Bryum argenteum*) and fire moss (*Ceratodon purpureus*).[56] Although closely related to the Arthropoda, they now belong to their own phylum, the Tardigrada. Discovered in 1777 they have survived all five of the mass extinctions thus far recorded and exist everywhere on Earth, from the deepest seas to the Himalayas. Like lichens they can survive a fling in outer space and endure years of desiccation. In one account tardigrades lumbered out of a piece of moss moistened after 120 years in storage.[57] When desiccated they retract their legs and shrink, becoming barrel-like 'tuns', able to survive for years in this form.[58] Tuns can reanimate after being chilled

A moss-and-lichen garden on a rock near Deception Pass, Washington, includes the hoary fringe-mosses (*Racomitrium canescens*), the bristly haircap moss (*Polytrichum piliferum*) and the antlered jellyskin lichen (*Scytinium palmatum*).

as low as –184.4 degrees Celsius. Tardigrades and mites keep company with mosses and lichens on nunataks. Rotifers congregate in the concave parts of moss leaves. They have a particular affinity for sphagnum moss, where they find a perfect watery home in the dead water-storing hyaline cells.[59] Alpine rotifers may become red after feeding on bryophyte detritus of red-hued sphagnums. Nematodes, the most abundant animals on Earth, coil their way through water

The convoluted surfaces of lichens provide refuge for tardigrades.

films in moss clumps, moving upwards during night and rain and downwards during day and drought. They are as numerous as rotifers in mosses.

Biologists look to mosses and lichens as survivors of the next mass extinction and colonizers of outer space. Moss proved viable after circa 1,500 years frozen 3 metres deep in a moss bank on Signy Island in the Antarctic.[60] Samples of the elegant sunburst lichen (*Xanthoria elegans*) subjected to space conditions at the International Space Station for one and a half years as part of LIFE (Lichen and Fungi Experiment) proved viable upon return to Earth.[61]

Astrobiologists interested in the propagation of life in space via rocks (lithopanspermia) consider lichens perfect candidates for the job. However, mosses and lichens have invaluable ecosystem functions here on Earth now as cosmopolitan extremophiles.

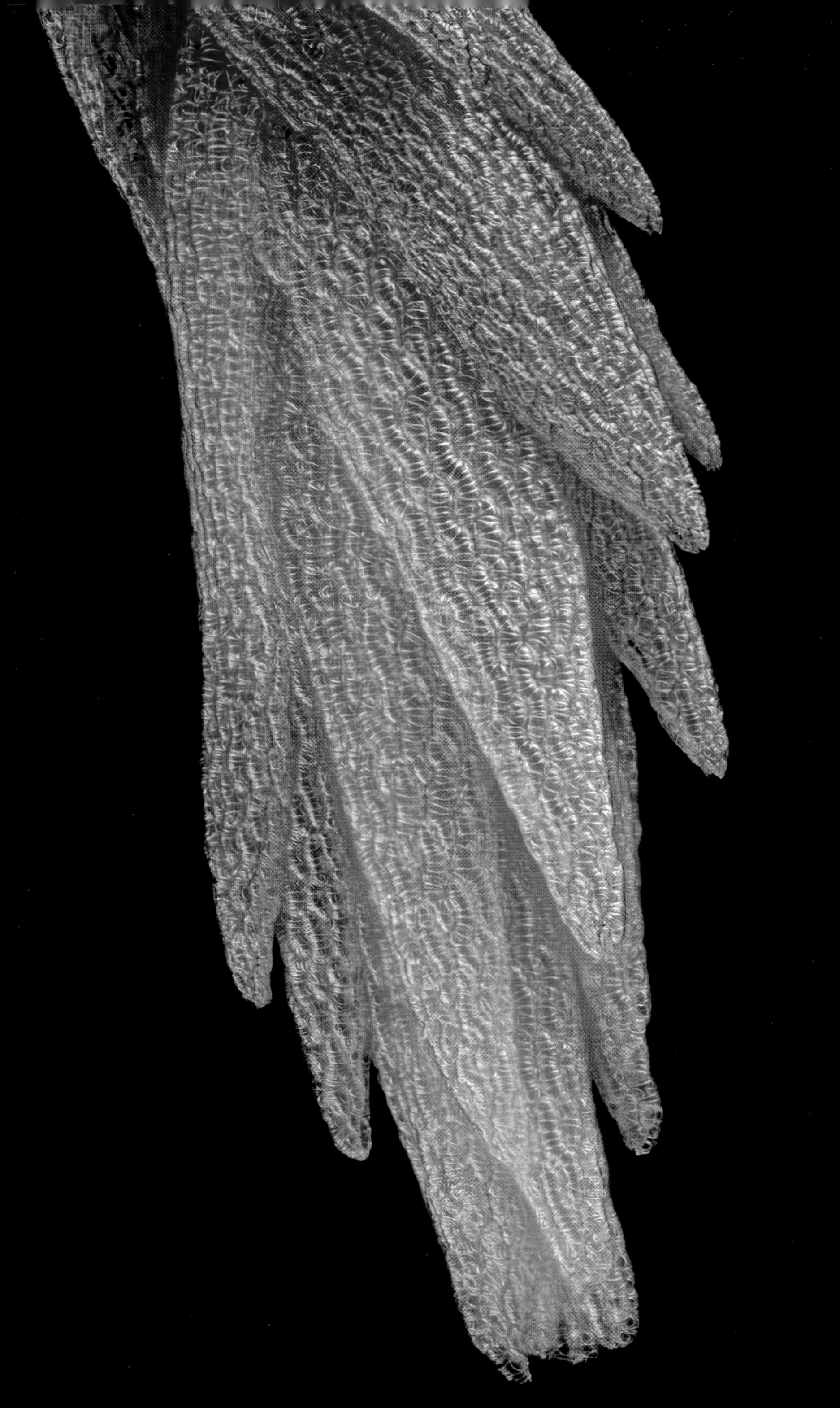

five Bogland

The quiet beauty of a bog conceals a contentious environment . . . A bog is a stressed ecosystem where few plants survive. The strategies by which they eke out an existence include, above all, a frugal use of nutrients. Death and decay release nutrients to be used again, to be cycled and recycled.

HOWARD CRUM, *A Focus on Peatlands and Peat Mosses* (1988)[1]

Peat is harvested from bogs, watery mires where the earth yawns open. The bottom is a breathless place – cold, acidic, anaerobic – with no oxygen to decompose the willow branches or the small, stiff faces of the foxes interred there. Sphagnum mosses wrap around fur, wood, skin, casting their spell of chemical protections, preserving them whole. Growth is impossible, and Death cannot complete her lean work.

KAREN RUSSELL, 'The Bog Girl' (2019)[2]

After a trip to Lapland in 1732, Linnaeus wrote, 'The whole of this Lapp country was bog, which is why I call it the Styx. No priest has ever painted Hell so vile that this does not exceed it, no poet described a Styx so foul that this does not eclipse it.'[3] Several hundred years later, bryologist Howard Crum (1922–2002) would spend many years researching sphagnum's lean lifestyle, describing the ecology of 'life in a wet desert'

The fibrils that strengthen water-storing cell walls are visible in this dehydrated shoot of *Sphagnum rubellum*.

in detail. A 2019 headline in *The Guardian* asked, 'Is Sphagnum the Most Underrated Plant on Earth?'[4] Similarly, a 2022 headline from the *New York Times* asked, 'Who Will Profit from Saving Scotland's Bogs?' The journalist noted, 'Until recently, peat bogs were derided as useless swampland – ideal, perhaps, for burying bodies in a murder mystery, but not much else.'[5] The newspaper has now made peatlands a topic in its Headway series. That sphagnum moss, also known as peat moss, is both underrated and extremely important relates to its remote, treacherous habitats and the recognition that this genus continues to sequester more carbon than any other plant on Earth.[6] Equally important, when damaged, peatlands release carbon into the atmosphere. Clearly, the time has come to admire sphagnum moss, the builder of bogs, and many writers are doing so now.[7]

Our current boreal peatlands are thought to have begun forming for the most part after the last Ice Age, 12,000 to 10,000 years ago, when deglaciation and flooding led to the water table rising, creating swamps, a process known as paludification (Latin *palude*, swamp). Sphagnum mosses belong to the Sphagnopsida, one of the earliest diverging lineages of mosses. A fossil bearing clear resemblance to sphagnum has been dated to the Ordovician era (485.4–443.8 mya) and is at this point the earliest land plant macrofossil.[8] It is thought that sphagnum species diversified in northern regions and later dispersed to mountain tops in Asia and Africa and the tropics, which suggests that they might be preadapted for climate change. Their ability to actively build and control boreal peatlands has earned them the title of ecosystem engineers and habitat manipulators.[9] In any event, ecologists remind us 'that Sphagnum mosses are not simply "adapted" or "tolerant" of a certain habitat – they create the habitat themselves. To apply traditional niche theory is therefore problematic.'[10] In other words, sphagnum creates its own ecological theory and practice.

Etymologists trace the roots of the word *sphagnum* to both modern Latin for '*sphagnos*, a kind of lichen' and Greek for '*sphagnos*, a spiny shrub, a kind of moss'.[11] Wading into bog-related terminology tests the limits of explanatory, descriptive language. In 'A Botanist in Swedish Lapland' (2017) James Prosek writes in his discussion of Linnaeus' 'compulsive' naming of plants that 'Nature is one interconnected system, but language necessitates that we chop it up and label the pieces, giving a false impression of what it's really like.'[12] It is wise to keep this in mind when trying to clarify terminology about peatlands. The International Peatland Society lists over 7,000 terms relating to peat and peatlands.[13] The profusion of words underscores the complexity of wetlands. Some names are familiar, such as bog, fen, marsh, mere, mire, moor, peatland and swamp, while others, such as carr, drunken forest, flow, flush, lagg, muskeg, pocosin, slough and chupaderas (sucking swamps), are less common but equally evocative of the often-alarming nature of these watery habitats.[14] Naming is particularly elusive here as these waterlogged places are always in transition, partly because of ecological succession and partly because of changes in weather cycles. Experts make murky, qualifying statements such as 'Quaking bogs are actually fens in some cases,' or, with reference to sphagnum, 'In this way it initiates the changes that

The leaves of the spiky bog-moss (*Sphagnum squarrosum*) spread in all directions (squarrose).

transform fens into open bogs and eventually spruce muskegs.'[15] The taxonomy of bogs includes blanket bogs, eccentric bogs, level bogs, string bogs, raised bogs and more.

Though the term 'peat bog' is usually reserved for sphagnum-dominated bogs, peat itself is not always synonymous with sphagnum moss. Peatland is an umbrella designation for wetlands that accumulate layers of decayed plant material, or peat.[16] Two new tropical non-sphagnum peatlands were found in 2014 – one the size of England in the Congo Basin and one of almost 35,000 square kilometres in Peru, part of the Amazon River Tributary. In these tropical peatlands, as in those of Indonesia, the formation of peat is due to the accumulation of the decay-resistant leaves of various tree species falling into preservative waters.[17] Sphagnum does, however, occur in the tropics and subtropics, and there are extensive true sphagnum bogs in Peru, Argentina and Chile. Most swamps and marshes do not produce peat, though they may be heavily clothed with mosses, such as the famous Moss Swamp of Romania, which is described as the little brother of the Great Moss Swamp of New Zealand.[18] 'Mos' or 'moss' is an old Scottish word for bog and was used in many place names that remain on maps today.

Apprehension about the nature and 'purpose' of bogs has undoubtedly led to prejudice against them. Even in the early nineteenth century people still believed that living near fens and bogs was unhealthy for humans, though in 1753 Robert Maxwell of Arkland addresses his readers cheerily:

> You seem to think that mosses contribute to the insalubrity of the atmosphere. I am naturally led to agree with you in thinking that the damp vapours that arise from them must tend to the unwholesomeness of our climate in general; yet it appears from experience, that people who inhabit mosses and mossy grounds, more especially in the higher parts of Britain, or where the peat-waters are incorruptible or antiseptic, are very healthy.[19]

Scottish painter and political satirist Isaac Cruikshank drew on bog lore for this 1796 caricature of John Bull.

Because the 'vapours' of swamps and marshes were thought to be the cause of 'marsh ague', many such wetlands were drained. Now we know that malaria, present throughout the ancient world, caused the ague. Strange lights and emanations from bogs, fens and mires, called will o' the wisps, became the stuff of local legend. The lights are phosphorescence caused by the spontaneous chemical ignition of peat gases such as methane and phosphine, which are products of decaying plants exposed to air in peatlands.[20]

Bogs are not bottomless, despite fears to the contrary. Reports of depth range from 13 to 50 metres.[21] But the deceptive appearance of a stable surface has led some travellers to fatal outcomes. In 1921 botanist J. W. Hotson described a wartime incident: 'It was through one of the open bogs of Poland that a German officer ordered his men to attack the retreating Russians, when practically the whole regiment sank out of sight.'[22] Other catastrophes include bog-bursts or bog-slides, frequent in Ireland since first described in 1697.[23]

Silver Flowe, a series of patterned blanket mires, is part of the Galloway and Southern Ayrshire UNESCO Biosphere Reserve in southern Scotland.

San Francisco's *The Call* of 31 December 1904 carried the headline 'Moving Bog Wipes Out a Village in Roscommon, Ireland'.[24] It describes how the Cloonshiever bog moved more than a kilometre in just a few days. In the UK, the word *flowe* is used to describe a flat extent of bogland. A group of bogs collectively known as the Silver Flowe, which occur in a valley called the Cauldron of the Dungeon deep in the Galloway Hills of Scotland, are in the care of Crichton Carbon Centre, an environmental charity that does peatland restoration training. A vast blanket bog in northern Scotland comprising 400,000 acres in Caithness and Sutherland counties has been proposed as a World Heritage site by the Flow Country, an organization promoting peatland restoration and conservation through a Green Finance Initiative.[25]

Sphagnum covers more of the world's land surface than any other single plant and is particularly dominant in the boreal peatlands of northern latitudes.[26] Though it grows, on average, only 1 millimetre per year, the ability of these small plants to make a huge impact is documented in the statistics: northern peatlands hold nearly 30 per cent of all terrestrial carbon and store twice as much carbon as all the world's forests.[27] Lists of bogs by country show broad representation throughout the world.[28] Some lists rank bogs by beauty.[29] Peat was first used as fuel in the seventh century. Draining bogs for farming and mechanical harvesting of peat have diminished them so drastically that many are now protected by organizations such as the Irish Peatland Conservation Council. Climate change and continued anthropogenic development threaten what remains. It was reported in 2005 that melting of the world's largest frozen peat bog, the size of France and Germany combined, in western Siberia had begun, bringing with it fears of the release of huge amounts of methane, a greenhouse gas much more potent than CO_2.[30] Clara Bog in Ireland began to receive protection in the 1980s with the relocation

Ireland's Clara Bog, now a nature reserve, symbolizes what Seamus Heaney celebrates in his poem 'Bogland' as 'unfenced country'.

of turf cutters. In 2019 it was the subject of a photographic essay by Emily Toner, a Fulbright–National Geographic Digital Storytelling fellow, titled 'The Secret World of Life (and Death) in Ireland's Peat Bogs' on the *New York Times* website. She included images of the oldest human fingerprint, which belonged to Old Croghan Man, Kerry bog ponies and lichen and sphagnum from the bog held on an outstretched hand.[31] Like Ireland, Finland is said to be half bog, while bogs make up one-fifth of Estonia. Vast sphagnum peatlands also occur in the Magellanic moorlands of Patagonia (Chile) and Argentina, with others in New Zealand and Tasmania.

The peculiar leaf architecture and unusual chemistry of sphagnum mosses are crucial in managing their waterlogged lifestyle.[32] The sphagnum leaf, like those of most other mosses, is only one cell thick but is distinctive in having two kinds of cells arranged in a striking and characteristic mesh-like pattern. Slender elongated cells filled with chloroplasts border much larger, 'broadly rhomboidal' hyaline cells. Hyaline, meaning clear or transparent, aptly describes their appearance. Dead at maturity, their function is to store water. Pores opening to both sides of the leaf allow entrance of water, which flows into the tiny pores by capillary action but does not exit unless pressure is applied. Hoop-like bands of strengthening material circle the cell walls. The enormous water-storing capacity and absorbency of sphagnum is said to be as much as 27 per cent of its dry weight.[33] Further, the stems have special retort cells (named after the vessels that medieval alchemists used for distillation) that also store water and shelter a host of microorganisms.[34] When multiplied millions of times, the minute leaves achieve considerable mass.

Sphagnologists use words such as fat, stout and tousled to describe the overall morphology of sphagnum. Growth occurs at the tip of each gametophyte, where a small cluster of cells actively divides. Below this zone the only growth is the elongation of these cells; there is no further addition of cells. The 'head', which contains the small cluster of actively dividing (meristematic) cells, is called a capitulum. It is surrounded by a tuft of little branches. As the stem

The lustrous bog-moss *Sphagnum subnitens* (var. *subnitens*) is said to be untidy and top-heavy.

elongates beneath the capitulum, these branches, which occur in groups called fascicles, mature into two forms – pendant branches that hang down, tangling around both each other and the stem for support and creating a wick for the upward transport of water, and lateral branches that spread horizontally to enhance capture of light for photosynthesis. Each sphagnum gametophyte is floppy, unable to stand alone. The growing tip of an individual sphagnum plant may maintain connection with the dead portion as far as 15 centimetres away. This entangling system of branches furthers the capture of rainwater from above and the uptake of water from below the bog's surface through capillarity. Radiocarbon dating and genetic analysis of peat cores have shown that sphagnum clones persist for on average more than four hundred years, and even more than 1,600 years in one case. Håkan Rydin and John K. Jeglum, authors of *The Biology*

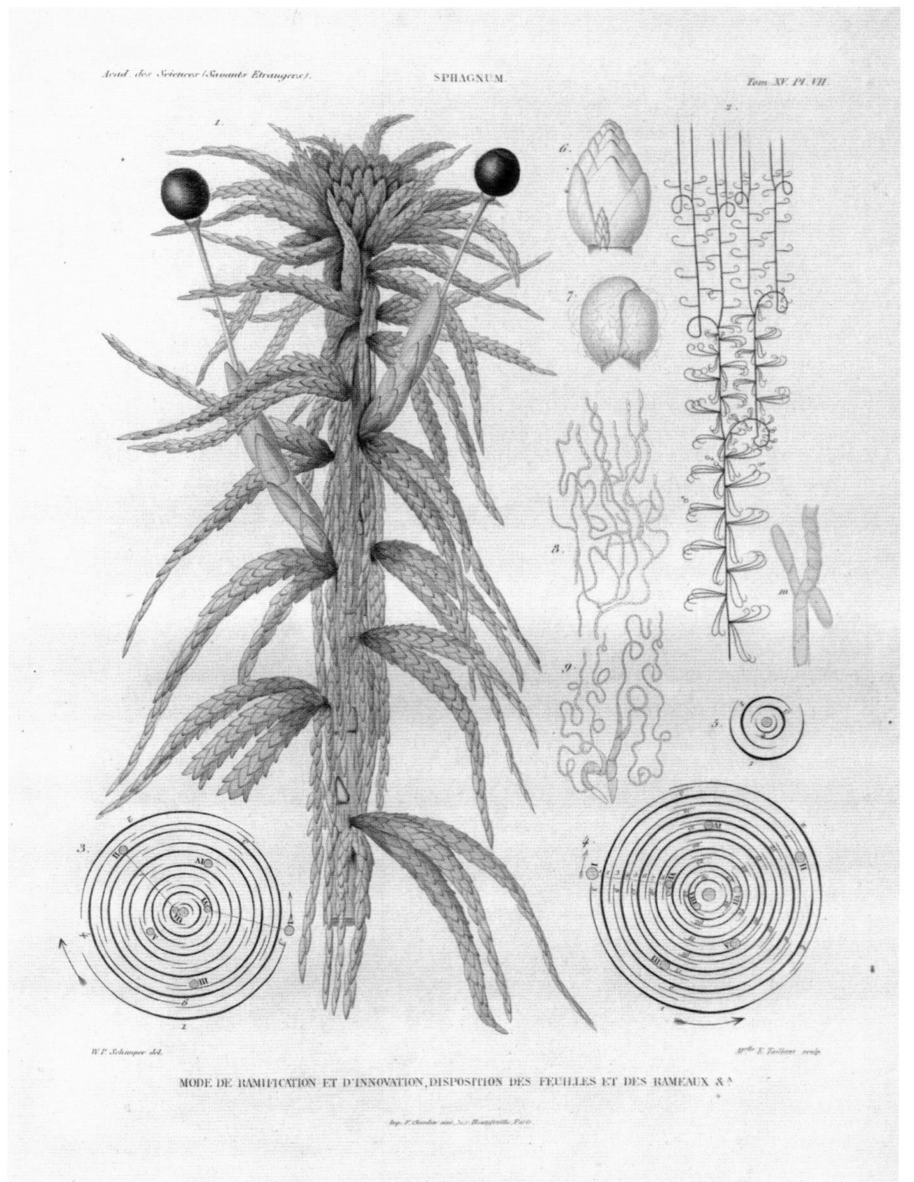

Sphagnum mosses have both horizontal and pendant branches that together facilitate waterlogging, as in this illustration from W. P. Schimper's *A History of Development of Peat Mosses* (1858).

of Peatlands, speculate that today we may be seeing 'eternally young' sphagnum plants that are genetically identical with those individuals from thousands of years ago.[35]

Sphagnum gametophytes support each other by massing together in extensive expanses called lawns. Though they are perhaps inelegant or unkempt in form when examined singly, together they create marvellous lawns of the billowy, floating variety, strong enough to support the weight of a moose or a bear, but not a regiment of soldiers, as we have seen. They often show variation into hummocks and hollows, created by specialist sphagnum species said to be hummock-forming or hollow-forming. These then give rise to larger patterns. Each species is specialized to take advantage of their location, either at the top of the hummock or in the dip of the hollow. They are constantly in transition, hummocks becoming hollows and hollows becoming hummocks. New data indicate that sphagnum gametophytes grow rhythmically according to temperature and lunar cycles, creating a uniform surface that reduces evaporation.[36] Thus, like the tiny terrestrial cushion mosses, the vast sphagnum lawn acts like a single individual, a 'supraorganismic system'.[37]

The colours, which are luminous in the bog atmosphere – ranging from apple green to gold, golden green and chartreuse in spring to peach, orange, red, scarlet and purple in late summer – invite closer inspection. Bryologist Robert Braithwaite (1824–1917) warned of the danger in 1880. He observes:

> Few persons can have traversed our moorlands without having had their attention attracted to the great masses of Sphagnum which adorn their surface – now in dense cushions of lively red – now covering some shallow pool with a vast sheet of light green, inviting it may be by its bright colour, but woe betide the inexperienced collector who sets foot thereon, for the spongy mass may be many feet in depth, and he may run the chance of never reaching terra firma again.[38]

The colour of the Skye bog-moss (*Sphagnum skyense*) varies impressionistically in each capitulum. Barkeval, a summit of the Rum Cuillin, Isle of Rum, Scotland.

Perhaps this is why Braithwaite, a prominent British bryologist who was known for his classic work *The Sphagnaceae or Peat-Mosses of Europe and North America*, was also known for avoiding fieldwork. His obituary lauds him as the sole finder of *Trematodon ambiguous* in Britain. Though known to be cosmopolitan, the ambiguous trematodon, as it is commonly known, is considered a 'vagrant' that has a habit of turning up here and there and then disappearing.[39] Researchers classify this type of life history as 'a short-lived shuttle'. A recent reoccurrence in France has been under surveillance for many years.[40]

Colour can be helpful in identification. Sphagnum taxonomy is challenging, however, because the various species are so 'plastic', a characteristic of plants in general, but one that is especially pronounced in sphagnum. They form species complexes and can look very different in even slightly dissimilar conditions. Bryologists have settled on saying that there are between 100 and 380 species of sphagnum. It is also known for ejecting spores explosively. In

The blushing bog-moss (*Sphagnum molle*) occurs in tightly packed velvety mats that are pale whitish-green with a hint of pink.

Structural Diversity of Bryophytes (2001), Crum includes a description of *S. capillifolium* spore explosion by Russian botanist Sergei Gavrilovich Navashin (1857–1930):

> The day was clear, and throughout the bog there was a continuous noise that I took to be the sound of gas bubbles bursting from water surfaces. However, to my great astonishment, I soon became aware of little reddish clouds puffing above most of the hummocks and noticed a noise associated with each puff. The continuous noise that I had attributed to inanimate nature was actually produced as a

> volley from many bursting capsules . . . Many times when bending over for closer examination, I felt the explosively discharged capsule lids strike my face.[41]

We know now that sphagnum generates long-distance spore dispersal – relatively speaking, for such short plants – through vortex rings, a means of getting spores high enough to enter turbulent air currents.[42] Sphagnum capsules are toothless, that is, lacking peristomes, but are unusually round: when the capsule wall shrinks, the lid is forced off explosively. The vortex ring is a self-sustaining force field that elevates a mushroom cloud of spores as high as 17 centimetres at speeds of 144 kilometres per hour.[43] Vortex rings occur in the animal kingdom in squid, jellyfish and the human heart. Explosive spore discharge may explain why small populations of sphagnum are found scattered haphazardly in unexpected places. Sphagnum spores are practically indestructible, resistant to both desiccation and ultraviolet radiation, and remain viable for decades and longer in some cases.

The Sphagnum Lifestyle: 'Smothering and Mothering'

Peat-forming sphagnum bogs live in constant existential crisis because of nutrient deficiency. While so many mosses are challenged by lack of water, sphagnum faces the opposite problem. They live high in the water table and are constantly waterlogged, with no input from soil nutrients. Although sphagnum is adept at recycling nutrients from its own decayed remains, the only new input comes from precipitation. Thus, bogs are said to be cloud-fed, or ombrotrophic (*ombros*, Greek for cloud + *trephein*, Greek for fed). Raindrops contain small amounts of nitrate, the most available form of nitrogen. Rain also causes flushing events that circulate those nutrients that can be recycled from below to the growing tips of sphagnum. But its survival is made difficult by its own acidifying lifestyle. Sphagnum transforms 'rich' (minerotrophic) fens into 'poor' (ombrotrophic) bogs. Fens are nutrient-rich because they receive nutrients from both the atmosphere and groundwater. They are relatively shallow and produce little peat. Bogs are nutrient-poor because sphagnum floats, with no connection to groundwater, which contains nutrients from the soil. Many feet below, peat accumulates to great depths in acidic, anoxic (low in oxygen) conditions where organisms of decay cannot work. Crum describes peat as 'a nutrient lock-up'.[44] Through the very nature of a waterlogged environment, various chemical means and its own growth patterns, sphagnum creates an environment that is acidic, low-nutrient and anoxic.

Unlike mosses that halt metabolism when stressed, sphagnum must keep growing in order to maintain its chemical control of the environment, because the growing tips have the greatest cation exchange capacity (CEC). CEC is a kind of chemical sleight of hand that allows sphagnum to accumulate valuable nutrient cations by releasing less valuable ones in an exchange system. Sphagnum cell walls contain uronic acids (also called sphagnan), up to 30 per cent of their dry mass, which endow sphagnum with a high CEC. In bogs, H^+ ions attached to the carboxyl group of the uronic acids are exchanged for nutrient cations like NH_4^+, Ca^{2+}, Mg^{2+}, and K^+. Sphagnum also releases whole organic acids, which increase acidity, and phenolics, which inhibit decay. Further, the decomposition of sphagnum tissue directly below the actively growing area leads to the release of humic acids. In this way, bogs become 100,000 times more acidic than fens.[45]

It is important to acknowledge the work of another group of mosses: feather mosses in the family Amblystegiaceae, in peatlands. Often called the brown mosses, they begin the process of succession that takes place ceaselessly in these watery habitats. They are said to 'initialize' terrestrialization – that is, they turn open water bodies into minerotrophic fens, but are outcompeted by sphagnum in bogs, partly because of the high acidity and partly because of the lower CEC. Brown mosses include genera such as *Drepanocladus*, *Hamatocaulis*, *Warnstorfia*, *Meesia*, *Campylium*, *Calliergon* and *Scorpidium*. *Amblystegium serpens*, also known as creeping feather moss or nano, is sold for use in the aquarium trade, though it doesn't grow submerged in water in nature. *Calliergon giganteum* gametophytes, which resemble miniature spruce trees, insulate vast areas of the Arctic tundra.

A bog is described as having two layers or zones: the acrotelm, the living portion where growth and some decay take place (the top 45 centimetres), and the catotelm, the anaerobic portion where peat accumulates for many feet owing to lack of decay. As sphagnum grows in the topmost few centimetres of the acrotelm, the capitulum (head) constantly generates new H^+ ions for exchange, receiving nutrient cations from precipitation and from cations rising in solution from decaying portions of sphagnum below in the acrotelm. It is an amazing feat of recycling in that the growing tip benefits from nutrients released by deceased portions of the same plant below and brought up by capillary flow through sphagnum's interlocking branches by flushing events. Sphagnum is particularly good at scavenging for

Left: Stunted, lichen-encrusted tamarack (*Larix laricina*) trees surround the periphery of Jam Pond, a notable bog in central New York.

Right: Insectivorous sundews (*Drosera* sp.) thrive in waterlogged, sphagnum-dominated bog communities.

nitrogen, though even this is tricky. Maintaining a low-nitrogen environment is important both for basic growth and in order to make the bog inhospitable for higher plants; too much nitrogen is dangerous. Researchers describe sphagnum mosses as 'masters of efficient N-uptake while avoiding intoxication'.[46]

Though sphagnum excludes the highest of higher plants, a stunning assemblage of small plants and shrubs tolerates the acidic, low-nutrient conditions, such as bog rosemary, bog laurel, Labrador tea, black crowberry, hackberry, leatherleaf, tamarack and baked apple berry. These acidophiles include a number of carnivorous and insectivorous species such as pitcher plants (*Sarracenia*), sundews (*Drosera*), butterworts (*Pinguicula*) and Venus fly traps (*Dionaea*), which are quite beautiful and even considered glamorous now but initially contributed to the 'unnatural' reputation of bogs. Several reindeer lichens, such as *Cladina* and *Cladonia*, colonize the surface of mature

hummocks.[47] Small acidophilic shrubs that have nitrogen-fixing nodules such as alder and sweet gale also contribute nitrogen, as do plentiful 'microbial associates', bacteria and archaea that live in the dead hyaline cells. Of these, methanotrophs metabolize methane, a potent greenhouse gas, and diazotrophs convert nitrogen from the atmosphere into a usable form. Additional nutrient inputs from the atmosphere include agricultural dusts (for example, calcium-rich dust from the Dakota wheat fields that blows far afield) and leaves blown from adjacent woodlands. Life in a bog is a group effort.

Crum summarizes the situation: bog organisms 'eke out a living by cation exchange, active uptake, internal translocation, symbiotic interdependence, carnivory, and parasitism. The effective use of atmospheric input, scant as it is, makes possible the continued development from fen to bog.'[48] In other words, rain is essential. Researchers are trying to anticipate how sphagnum will adapt as key variables like precipitation, temperature and water table depth change. A number of online field trips portray the diversity of bog communities.[49] Crum describes sphagnum's approach as 'smothering and mothering' and cites a passage written by Wilhelm Philippe Schimper (1808–1880), the influential bryologist responsible for the six-volume *Bryologia europaea* (1836–55), as his inspiration for the phrasing:

> For just as the Sphagna suck up the atmosphere and convey it to the earth, do they also contribute to it by pumping up to the surface of the tufts formed by them the standing water which was their cradle, diminish it by promoting evaporation and finally also by their own detritus, and by that of the numerous other bog-plants to which they serve as a support, remove it entirely, and thus bring about their own destruction.[50]

This description beautifully captures the way that the lean work of death and the lean work of life balance in a bog. Sphagnum moss

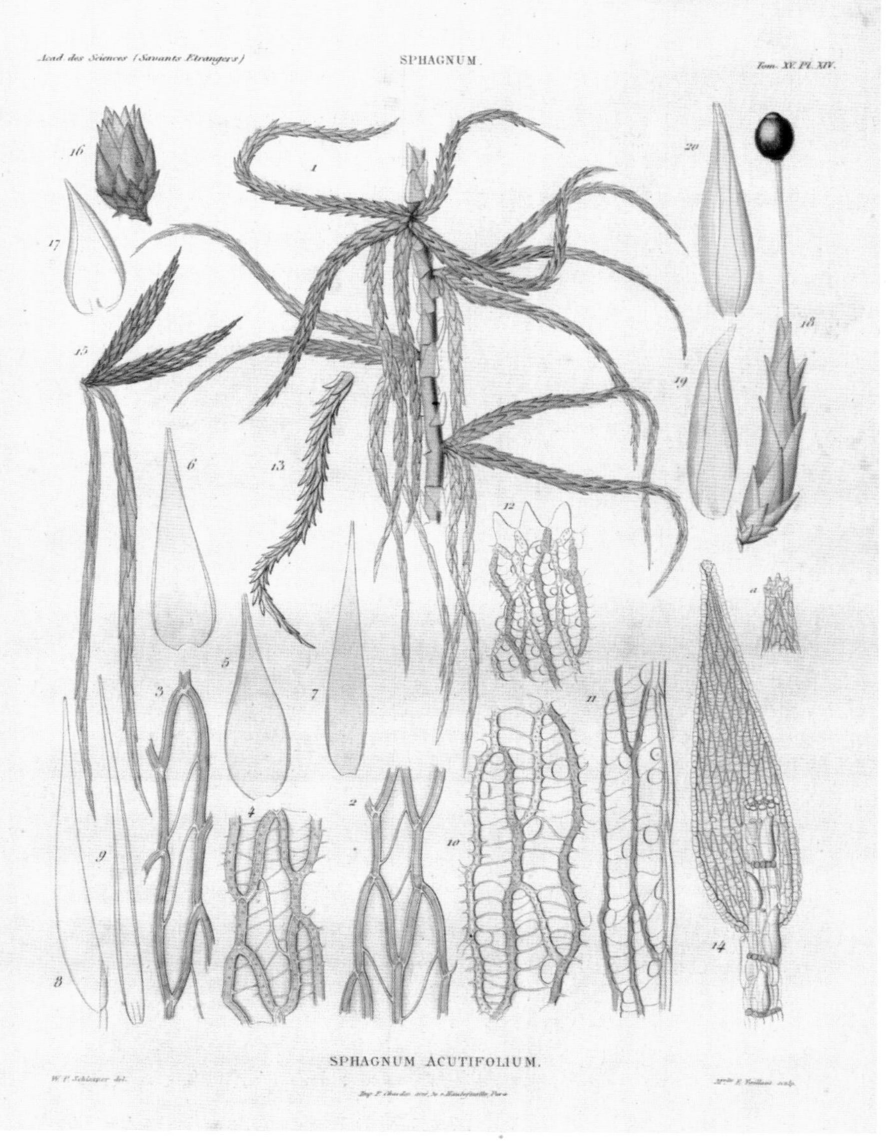

W. P. Schimper's illustrations in *A History of Development of Peat Mosses* (1858) capture the diversity and beauty of sphagnum species.

is now a key subject of environmental projects that encourage innovation in sustainable use and restoration of natural resources, and scientific studies that hope to understand the effects of climate change. Bogs are no longer regarded as malign, unproductive landscapes in need of human intervention. Though the continual unearthing of bog bodies tanned in the peaty waters perpetuates our sense of unease about how they archive violent death, bogs have stored homely household items such as kegs of butter as well.

Recently the UK Government announced a total ban on horticultural uses of peat moss by 2024, and the United States will follow by 2030. In the 1990s two organizations with competing interests, the International Peatland Society, which favours commercial 'carrier functions', and the International Mire Conservation Group, which favours 'non-material life-support functions', met to prepare a document that would represent a cooperative 'wise use' approach to peatlands; it was published in 2002.[51] The authors note in the first paragraph that 'everybody wants something from peatlands'; many or most will be disappointed. The Scottish Government has promised to reimburse 80 per cent of the amount spent on repair of boglands by landowners, who will be entitled to keep the profits from carbon-credit sales if their repairs are certified.[52] In answer to the questions posed at the beginning of this chapter, it is fair to say that sphagnum is no longer underrated and that a profit motive can be generated for the non-use and restoration of boglands.

six

Literary Ecology

But good-bye, good-bye, old mast-head! What's this? – green? Aye, tiny mosses in these warped cracks. No such green weather stains on Ahab's head!

HERMAN MELVILLE, *Moby Dick* (1851)[1]

Let us be clear about this at least: Slough House is not in Slough, nor is it a house . . . The front door, as stated, lurks in a recess. Its ancient black paintwork is spattered with road splash, and the shallow pane of glass above its jamb betrays no light within. An ancient milk bottle has stood in its shadow so long, city lichen has bonded it to the pavement.

MICK HERRON, *Slow Horses* (2010)[2]

Captain Ahab speaks these words in the last moments of his life. It is the third day of the chase, and he is about to go down with the white whale, a truly remarkable moment for Melville to introduce a tiny moss. It seems unlikely that it was a random act of writing, rather that this is how Melville chooses to show Ahab judging himself, seeing the scale of his mania in comparison with the tiny moss finding home in a warped crack on the deck of a whaling ship, patient with a small portion of life. Likewise, inept MI5 agent River Cartwright in *Slow Horses* will end up bonded by lichen to the pavement for a lifetime if he doesn't get his act together.

Voyaging through literature while being attentive to mosses and lichens is much like finding them in real life – here, there and everywhere. Numerous poems of the Victorian era praised mosses and lichens as adornments of rocks and trees, with gratitude for nature's artistry. Now that their biology is better known, there is more to say. There are many literatures in which to search for mosses and lichens – in scientific journals, novels, poems and general non-fiction. The use of specialist vocabulary to report recent research discoveries poses a challenge for the general reader. But literary texts can act as a microscope to help us notice mosses and lichens in different contexts.

Sometimes a scientist reaches out to the general audience, as Robin Wall Kimmerer does in *Gathering Moss: A Natural and Cultural History of Mosses* (2003).[3] Kimmerer, a professor at the State University of New York college of environmental science and forestry, has become a treasured advocate for the natural world. Her book is a series of personal essays in which she shares her deep knowledge of mosses gained during a career devoted to moss ecology. It won the John Burroughs Medal and has brought generations of people closer to mosses. Elizabeth Gilbert made a pilgrimage to see Kimmerer when she was researching her book *The Signature of All Things*.[4] She describes in her blog post 'Me and My Moss Mentor!' how she showed Kimmerer her seventy-page outline, developed over three years of research, because, since mosses figure heavily in the plot, she needed to know that she had interpreted their story correctly.[5] Kimmerer gave her confirmation. Gilbert said that she set out to write a Dickensian novel and was inspired by finding her great-grandfather's copy of the 1784 edition of *Cook's Voyages* in her attic.[6] *The Signature of all Things* is a bold, capacious novel with an exuberant plot. It tells the story of a young woman who begins the study of mosses at White Acres, her father's estate, in nineteenth-century Philadelphia. After a life of turmoil and travail she begins writing

> the history of the Moss Wars of White Acre. She wrote the story of the twenty-six years she had spent observing the advance and retreat of competing colonies of moss across the tumble of boulders at the edge of the woods. She focused her attention most specifically upon the genus

> Dicranum, because it demonstrated the most elaborate range of variation within the moss family. Alma knew of Dicranum species that were short and plain, and others that were dressed in exotic fringe. There were species that were straight-leaved, others that were twisted . . .[7]

The genus *Dicranum* has more than three hundred species.[8] The windswept broom moss (*D. scoparium*) is common in the Northern Hemisphere. It is noticeable, tall at 10 centimetres and distinctive for its sideways-leaning coiffure.

In 1992 the British Bryological Society published a special volume titled *Mosses in English Literature* to investigate the 'literary ecology' of mosses and take note of how the views of non-bryologist observers have changed through the centuries.[9] It is a collection of almost three hundred quotations, arranged alphabetically by author, covering 2,000 years of writing. The compiler of quotations, Sean R. Edwards, points out that 'there are evident general trends in the use of the word moss, though sadly it is clear that negative associations persist. Moss may still be said to have an image problem.'[10] One colleague, Eustace Jones, made the comment that 'It is remarkable that the early references associate moss with discomfort, disease,

Children's author Jan Brett portrays the beauty of mosses and lichens throughout *Mossy* (2012), her story about a turtle with a carapace garden.

Moss
by William Barnes

O rain-bred moss that now dost hide
The timber's bark and wet rock's side,
Upshining to the sun, between
The darksome storms, in lively green,
And wash'd by pearly rain drops clean,
Steal o'er my lonely path, and climb
My wall, dear child of silent time.
O winter moss, creep on, creep on,
And warn me of the time that's gone.

Green child of winter, born to take
Whate'er the hands of man forsake,
That makest dull, in rainy air,
His labour-brighten'd works; so fair
While newly left in summer's glare;
And stealest o'er the stone that keeps
His name in mem'ry where he sleeps.
O winter moss, creep on, creep on,
And warn us of the time that's gone.

Come lowly plant that lov'st, like me,
The shadow of the woodland tree,
And waterfall where echo mocks
The milkmaid's song by dripping rocks,
And sunny turf for roving flocks,
And ribby elms extending wide
Their roots within the hillock's side.
Come winter moss, creep on, creep on,
And warn me of the time that's gone.

Come, meet me wandering, and call
My mind to some green mould'ring hall
That once stood high, the fair-wall'd pride
Of hearts that lov'd, and hoped, and died,
Ere thou hadst climb'd around its side:
Where blooming faces once were gay
For eyes no more to know the day.
Come winter moss, creep on, creep on,
And warn me of the time that's gone.

While there in youth, — the sweetest part
Of life, — with joy-believing heart,
They liv'd their own dear days, all fraught
With incidents for after-thought
In later life, when fancy brought
The outline of some faded face
Again to its forsaken place.
Come winter moss, creep on, creep on,
And warn me of the time that's gone.

Come where thou climbedst, fresh and free,
The grass-beglooming apple-tree,
That, hardly shaken with my small
Boy's strength, with quiv'ring head, let fall
The apples we lik'd most of all,
Or elm I climb'd, with clasping legs,
To reach the crow's high-nested eggs.
Come winter moss, creep on, creep on,
And warn me of the time that's gone.

Or where I found thy yellow bed
Below the hill-borne fir-tree's head,
And heard the whistling east wind blow
Above, while wood-screen'd down below
I rambled in the spring-day's glow,
And watch'd the low-ear'd hares upspring
From cover, and the birds take wing.
Come winter moss, creep on, creep on,
And warn me of the time that's gone.

Or where the bluebells bent their tops
In windless shadows of the copse;
Or where the misty west wind blew
O'er primroses that peer'd out through
Thy bankside bed, and scatter'd dew
O'er grey spring grass I watch'd alone
Where thou hadst grown o'er some old stone.
Come winter moss, creep on, creep on,
And warn me of the time that's gone.

Polymath William Barnes (1801–1886) celebrated rural life in Dorset in both dialect and common English.

and death; mosses are not regarded as objects of admiration. The interesting question is when and why did the association become not merely romantic but also pleasurable.'[11] Certainly, beginning with Ovid (43 BCE–17/18 CE), mosses and lichens were associated with death (tombstones) and filth, followed by barrenness and stagnation owing to their presence in waste places, and thence to more abstract associations such as loneliness, solitude, shadows, dreams and haunting.

Their image improved as time went on. Edwards notes enthusiastically that 'Mosses positively flourished during the Romantic period [1798–1837]'.[12] Ann Radcliffe (1764–1823), popular author of Gothic fiction, loved travelling with her husband and their dog, Chance. During a tour of the lakes of Lancashire, Westmoreland and Cumberland in 1794 they found themselves walking below a fell called the Nab, where 'little streams of crystal clearness wandered silently'. She observed:

> The gray stones, that grew among the heath, were spotted with mosses of so fine a texture, that it was difficult to ascertain whether they were vegetable; their tints were a delicate pea-green and primrose, with a variety of colours, which it was not necessary to be a botanist to admire.[13]

Dorothy Wordsworth was also observant of mosses in her Grasmere journals and elsewhere. In her account of the tour that she, her brother and Samuel Taylor Coleridge made to Scotland in 1803, she describes a fog hut. 'Fog' was a Scottish word for moss, and fog huts were a feature of Scottish country estates at the time:

> We came to a pleasure-house, of which the little girl had the key; she said it was called the Fog House because it was lined with 'fog', namely moss . . . within [it] was like a hay stack scooped out. It was circular, with a dome-like roof, a seat all round fixed to the wall, and a table in the middle, – seat,

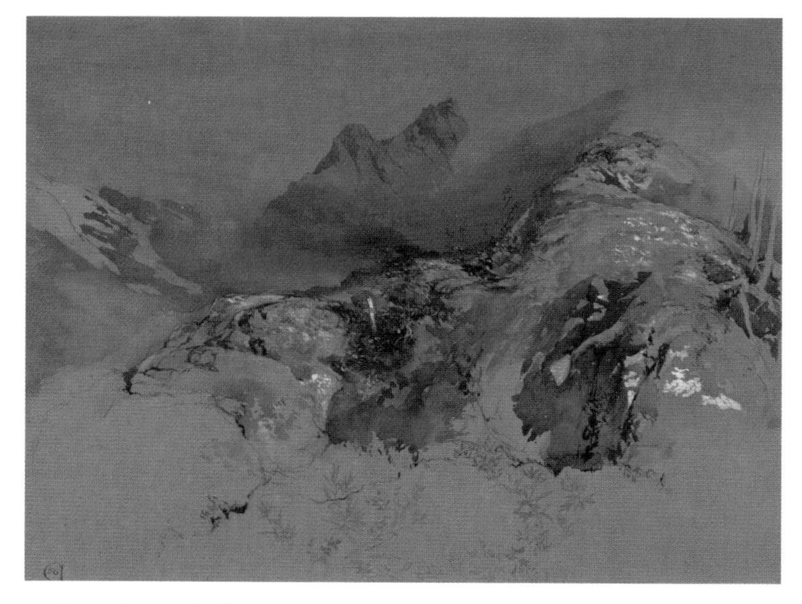

John Ruskin's *Study of Rocks and Lichens in the Glen below the Montets in the Ascent to Chamonix* highlights his 'telescopic' vision; 1849, graphite, ink and watercolour.

> wall, roof, and table all covered with moss in the neatest manner possible. It was as snug as a bird's nest.[14]

The Victorians were lovers of mosses and lichens, none more so than John Ruskin (1819–1900), who celebrated them in artwork and in prose. In 'Ruskin and Lichen' art historian Kate Flint discusses both in the context of the ecological violence caused by human civilization. She writes, 'Ruskin, in turning his attention to moss and lichen, uses – metaphorically speaking – both microscope and telescope. He sees them both in delicate detail and then responding to a setting sun at four or five miles' distance.'[15] Bloggers today praise Ruskin's word pictures about lichens and mosses, which show, as Flint suggests, his vision of the small writ large.[16]

Edwards laments that few of the quotes he gathers are humorous. But when we come to the Ws we find P. G. Wodehouse (1881–1975). Edwards quotes passages from *Blandings Castle* (1935) where Lord

Emsworth argues with his sister and his Scottish gardener, McAllister, about replacing a mossy path with gravel. Wodehouse used moss to describe one of his most unique characters, Gussie Fink-Nottle: 'This Fink-Nottle, you see, was one of those freaks you come across from time to time during life's journey, who can't stand London. He lived year in and year out, covered with moss, in a remote village in Lincolnshire.'[17]

A Rolling Stone Gathers No Moss

There are many versions of the well-known proverb 'a rolling stone gathers no moss', and interest in this seemingly straightforward saying has never flagged. While Publilius Syrus, a Roman playwright, was initially considered the original source (*c.* 42 BCE), currently Egbert of Liège in 1023 earns the first accepted documentation. However, its popularity began in 1500 with the publication of *Adagia*, a text by Erasmus, and has continued throughout succeeding centuries. Some see the moral as an admonishment to stay rooted, while others see it as an invitation to cast off fetters and explore new opportunities. Shakespeare projected the latter interpretation in *The Comedy of Errors* when he seems to disparage 'the idle moss'.[18] The most troubling use of the saying occurred in the 1950s when it was used as a tool for psychological analysis in American Veterans Affairs hospitals, particularly in cases of schizophrenia. A patient unable to restate the proverb in a metaphorical way was suspected of mental illness. This is chillingly portrayed in Ken Kesey's *One Flew over the Cuckoo's Nest* (1962). McMurphy, a rebellious patient, is forced into a lobotomy partly because of his literal rewording of the proverb.

J.R.R. Tolkien plays with the proverb at the end of *The Return of the King*. Gandalf says,

> I am going to have a long talk with Bombadil; such a talk as I have not had in all my time. He is a moss-gatherer, and I have been a stone doomed to rolling. But my rolling days are ending, and now we shall have much to say to one another.[19]

Members of Tolkien forums speculate on what was said in that long talk.

In an essay on Rudyard Kipling as a globe-trotter, G. K. Chesterton (1874–1936) wrote:

> In the heated idleness of youth we were all rather inclined to quarrel with the implication of that proverb which says that a rolling stone gathers no moss. We were inclined to ask, 'Who wants to gather moss, except silly old ladies?' But for all that we begin to perceive that the proverb is right. The rolling stone rolls echoing from rock to rock; but the rolling stone is dead. The moss is silent because the moss is alive.[20]

Apparently Chesterton loved turning popular sayings inside out.[21] George Bernard Shaw also offered convoluted thoughts about the saying in his treatise on child-rearing: 'We keep repeating that silly old proverb that a rolling stone gathers no moss, as if moss were a desirable parasite. What we mean is that a vagabond does not prosper . . . children can be thoroughly mobilized without making vagabonds of them.'[22] Perhaps the proverb is best taken literally.

Many writers fail to represent details of mosses in their work. Alexander Pope (1688–1744) was an exception. In addition to being a poet and philosopher, he was also a lover of the green world and a keen gardener, having grown up near Windsor Forest, where his appreciation of mosses may have begun.[23] In 1733, in Epistle I of his series of *Moral Essays in Epistles to Several Persons*, he wrote a lovely passage making a connection between variation in mosses and human moods:

> There's some peculiar in each leaf and grain,
> Some unmark'd fibre, or some varying vein:
> Shall only man be taken in the gross?
> Grant but as many sorts of mind as moss.
> That each from other differs, first confess;
> Next, that he varies from himself no less:

Add nature's, custom's, reason's, passion's strife,
And all opinion's colours cast on life.
Our depths who fathoms, or our shallows finds,
Quick whirls, and shifting eddies, of our minds?[24]

He makes a similar comparison between the numerous kinds of moss and the 'shifting eddies' of the human mind in *The Dunciad Variorum* (1729): 'The mind, in metaphysics at a loss/ May wander in a wilderness of moss.'[25] Pope placed a footnote to the word 'moss' that reads 'Of which the Naturalists count above three hundred species.' For

Timmy Tiptoes, the little grey squirrel in Beatrix Potter's *The Tale of Timmy Tiptoes*, lies imprisoned in a woodpecker's hole, enveloped in a bed of moss.

Pope, moss conjured a wilderness rather than a cozy confinement, perhaps because he had trained his eye to distinguish variations in the greenery he loved so much.

There are many allusions to mosses as providers of comfort, perhaps because they look so at home and comfortable, even on rocks – and because, as we have seen, mosses have medicinal properties. In several scenes in Madeline Miller's novel *Circe* (2018), moss has a role. When Circe takes her uncle, the flayed Prometheus, a goblet of nectar after the Fury's horrific whipping, she finds that his skin 'smelled of green moss drenched with rain'.[26] Later, retreating to a forest after her father, the sun god Helios, has blasted her with his radiance in anger, she lays on a bed of moss to cool and soothe her burns. In *The Tale of Timmy Tiptoes* (1911) Beatrix Potter writes, 'He found himself tucked up in a little moss bed, very much in the dark, feeling sore; it seemed to be underground.'[27] In Tove Jansson's *The Moomins and the Great Flood* (1945), her first book in the Moomintroll series, the tired adventurers – Moomintroll, Moominmamma and the little stranger – sleep on beds of moss after a perilous journey through the Great Swamp. Caveats apply, however: the amateur naturalist knows that moss cushions sometimes harbour stinging ants and the like.

While Tove Jansson allowed her sweet moomintrolls to sleep on mosses, she did not want humans to do the same. In her adult novel *The Summer Book*, a grandmother and her granddaughter, Sophia, spend a summer together on a small island in the Gulf of Finland. It is an unsentimental look at how grief, old age and relationships sort themselves out in a setting where natural history is as important as human history. The grandmother tells the granddaughter that:

> Only farmers and summer guests walk on the moss. What they don't know – and it cannot be repeated too often – is that moss is terribly frail. Step on it once and it rises the next time it rains. The second time, it doesn't rise back up. And the third time you step on moss, it dies.[28]

Indeed, trampling of the fragile moss-and-lichen groundcover of the Fjaðrárgljúfur canyon in Iceland by Justin Bieber fans, attracted by the music video 'I'll Show You', led to the canyon's closure. *The Summer Book* was first published in 1972. On its fiftieth anniversary and republication in 2022, celebratory articles appeared. Lucy Knight highlights this passage as central to the book. She writes, 'This attitude of care and preservation is at the heart of *The Summer Book*: it proposes that every plant, every insect – and, indeed, every person – has a right to exist and to be looked after. And, 50 years on, that message is more vital than ever.'[29]

Mosses often appear in fairy tales, the best known being 'Mossycoat'.[30] In order to save her daughter from a hawker, a long-widowed mother presents her with a magical coat of moss that she has been crafting for many years and renames her Mossycoat:

> Green moss and goold thread, dat's what it was med on; just dem two things. 'Mossycoat,' she called it . . . It was a magic coat, she said, a wishing coat, she telt her daughter; when she'd got it on, she telt her she'd only to wish to be somewhere, and she'd be dere dat were instant, and de same if she wanted to changer hersel' into summat else, like to be a swan or a bee.[31]

In Philip Pullman's version from 2009, the mossy coat is beautifully bryological:

> It was as green as a spring morning, as fresh and soft as a breeze out of the west. All the mosses her mother had gathered from pond and meadow and millstream over 18 years were bright and living yet: she'd plaited and woven them so cunningly that all the tiny moss-leaves were still alive . . . The mossy coat was so light and fine you could fold it all into a thimble, and yet so strong you couldn't tear it with your teeth.[32]

'The cup of healing': Jeermit offers a potion containing blood-red sphagnum moss to the ailing princess in John Duncan's illustration for the Scottish wonder tale 'The Princess of Land-under-Waves', from Donald A. Mackenzie, *Wonder Tales from Scottish Myth and Legend* (1917).

Angela Carter writes that fairy tales 'are all of them the most vital connection we have with the imaginations of the ordinary men and woman whose labour created our world'.[33] How reasonable for a storyteller to imagine that an impoverished widow might use easily available moss as a sewing material to create a magical garment.

'Blood-red' sphagnum moss is central to a Scottish wonder tale called 'The Princess of Land-under-Waves', anthologized by folklorist Donald A. Mackenzie (1873–1936) in 1917. Mackenzie writes in his introduction that:

> Those who know the west coast [of Scotland] will be familiar with the glorious transparency of the hill-surrounded lochs

Sphagnum pigments found in cell walls (sphagnorubins) influence decomposition and absorption of UV radiation (red peat moss, *Sphagnum rubellum*).

> in calm weather. When the old people saw the waters reflecting the mountains and forests, the bare cliffs and bright girths of green verdure, they imagined a 'Land-under-Waves' about which they, of course, made stories.[34]

The story concerns a princess who is fleeing the Dark Prince-of-Storm, who means to marry her and seize her father's kingdom. She arrives alone by boat to seek the help of Finn, one of the god-like Feans who live in Land-under-Waves. Finn solves her problem and promises to help the princess whenever she appeals to him, and so she does, a year and a day later, when she becomes very ill. Finn sends the healer Jeermit, 'the fairest of the Feans'. In a long, long walk over an unending plain he finds three clumps of red sphagnum moss, which he presents to the princess, saying, 'If you will take them in a drink, they will heal you, because they are the three life drops of your heart.'[35]

A Moss-and-Lichen Thread

A lively thread on X ensued when author and professor Robert Macfarlane (@RobGMacfarlane) posted a quote from Robin Wall Kimmerer's *Gathering Moss* on 24 February 2018:

> Mosses and other small beings issue an invitation to dwell for a time right at the limits of ordinary perception. Look in a certain way and a whole new world can be revealed. But a cursory glance will not do it.[36]

The post prompted a series of favourite moss passages from users of X. Maarten De Cock offered a quote from John Cowper Powys's *A Glastonbury Romance* (1932):

> But the odour which floated now over that little garden of Benedict Street . . . had yet another pervading element in it – the scent of moss. Not a patch of earth in any of those spinneys, and copses, and withy beds, that edged those water-meadows, not a plank, not a post, in the sluices and weirs and gates of those wide moors, but had its own growth, somewhere about it, of moss 'softer than sleep'. More delicately, more intricately fashioned than any grasses of the field, more subtle in texture than any seaweed of the sea, more thickly woven, and with a sort of intimate passionate patience, by the creative spirit, than any forest leaves or any lichen upon any tree trunk, this sacred moss of Somersetshire would remain as a perfectly satisfying symbol of life if all other vegetation were destroyed out of that country. There is a religious reticence in the nature of moss. It vaunts itself not, it proclaims not its beauty; its infinite variety of minute shapes is not apprehended until you survey it with concentrated care.[37]

Viewed as successor to Thomas Hardy in sensibility, John Cowper Powys mentioned a life-changing experience in his *Autobiography* (1934) upon seeing green moss and yellow stonecrop.

Powys placed his story in the landscape of Somerset and Dorset, beloved of Thomas Hardy, who similarly appreciated the moss of Dorset woodlands: 'Variety upon variety, dark green and pale green, mosses like little fir-trees, like plush, like malachite stars, like nothing on earth except moss.'[38]

Climate warrior Katie Holten entered the thread with praise of Max Porter's 'study of mosses and memories growing on a red brick wall' in *Grief Is the Thing with Feathers* (2016). Macfarlane then referenced the 'highly lichenous' thread on X of modernist Scottish poet Drew Milne, who describes himself on X as a 'poet, critic and dark ecologist of the neo-modernist tendency in solidarity with the biotariat (especially lichens) against the carbon liberation front'. A poem sequence called 'Lichens for Marxists' appears in Milne's *In Darkest Capital* (2017). The compositional fluidity of lichens makes them the progressives of the biological world.

French author Pascal Quignard mentions lichens and mosses frequently in his dreamlike contemporary novel *Mysterious Solidarities* (2019). The narrative follows Claire as she wanders the headland of the Côte d'Émeraude of Brittany, near Dinard, spying on her former, and perhaps occasionally current, lover Simon from her 'rocky shelter . . . her cheek against the hot granite, against the little golden yellow lichen'.[39] Mention of this lichen occurs repeatedly. There are pink lichens as well, and she watches 'a wonderful little snail' eat moss. As Claire ages, she becomes part of the landscape: 'she had begun to give off a light smell of sweat, hay, salt and sea air, of sea, granite, and lichen.'[40] She protects the special places that have become part of her: 'From the granite, the quartz, the seams of black dolerite, the pink sandstone, the lichen, the moss, the sand and the seaweed, she cleaned away all the paper bags, corks, bottle tops and other litter, the twigs, gulls' feathers, cigarette butts and dried seaweed, which she stuffed into the dirtiest backpack I've ever seen.'[41] Readers begin to realize that there are many 'mysterious solidarities' at hand, none stronger than that between Claire and the landscape of granite and lichen and moss. Quignard singles out lichens in this sentence: 'God is so old in the tiniest patch of lichen, in the fingernail that lifts it, in the round eye that approaches it – an eye that is the product of the sun.'[42] He mentions God twice in the book, both times in relation to lichens.

On 6 April 2018 the British Bryological Society tweeted, 'Anyone tempted to do a haiku with lichens?' While mosses and lichens appeared in many effusive Victorian poems, they have not received a more modern take until recently. Writer and translator Forrest Gander quotes Drew Milne at the beginning of *Twice Alive: An Ecology of Intimacies* (2019): 'a garden without lichens/ is a garden without hope'.[43] This sets the stage for a 'lichenous' collection of poems. Gander was inspired by fieldwork with Anne Pringle, who frequents graveyards in her studies of ageing in lichens.[44] Gander writes in the foreword to his book that it was collaboration with Pringle that led to his poetic theme of intimacy: 'Anne and other contemporary biologists are saying that our sense of the inevitability of death may be determined by our mammalian orientation. Perhaps some forms of life have "theoretical immortality".'[45] Dust-like fragments of lichens, complete with all symbiotic partners, disperse widely, establishing new populations – perhaps endlessly. British science fiction writer John Wyndham (1903–1969) made use of the link between lichens and longevity in his lively *Trouble with Lichen*. Gander also uses lichens as a model to suggest that human identity is 'combinatory'.[46] The title poem, 'Twice Alive', is dedicated to the lichen. It is full of lovely details about lichens, all the while honouring their biology. The first stanza reads: 'mycobiont just beginning to en-/ wrap photobiont, each to become/ something else, its own life and a/ contested mutuality, twice alive/ algal cells swaddled in clusters'.[47]

In the acknowledgements Gander thanks his 'international poet-companions among the lichens'. His list includes Drew Milne, mentioned above, Lew Welch (1926–1971?) and Brenda Hillman. Lew Welch was a member of the literary movement known as the Beat Generation that flourished in California after the Second World War. Welch's poem 'Springtime in the Rockies, Lichen' appeared as a broadside published by Cranium Press in 1971. In the poem Welch offers a plaintive query: 'In this big sky and all around me peaks &/ the melting glaciers, why am I made to/ kneel and peer at Tiny?'[48] Indeed, the world's many baffling changes in scale can unsettle our vision and thinking. In a talk given at Naropa University in 1980 about 'Lewie', as he called him, and the poem, Philip Whalen (1923–2002), friend, Beat poet and Zen monk, asked another question, 'But what is it that lichen are doing? They're sitting there very quietly growing, very slowly, and not bothering anybody. They don't even get involved in the whole bee and flower business.' He concludes that the poem is metaphorical, linking lichens with artists, who 'by being quiet and working slowly and turning purple, will last, will endure, will do something when everything else is gone'.[49] Certainly the poem can be taken metaphorically about human endeavour, but it seems likely that Welch was showing reverence for the lichen as a persevering life form, its 'art' being the crumbling of rock.

Springtime in the Rockies, Lichen

All these years I overlooked them in the
racket of the rest, this
symbiotic splash of plant and fungus feeding
on rock, on sun, a little moisture, air —
tiny acid-factories dissolving
salt from living rocks and
eating them.

Here they are, blooming!
Trail rock, talus and scree, all dusted with it:
rust, ivory, brilliant yellow-green, and
cliffs like murals!
Huge panels streaked and patched, quietly
with shooting-stars and lupine at the base.

Closer, with the glass, a city of cups!
Clumps of mushrooms and where do the
plants begin? Why are they doing this?
In this big sky and all around me peaks and
the melting glaciers, why am I made to
kneel and peer at Tiny?

These are the stamps on the final envelope.

How can the poisons reach them?
In such thin air, how can they care for the
loss of a million breaths?
What, possibly, could make their ground more bare?

Let it all die.

The hushed globe will wait and wait for
what is now so small and slow to
open it again.

As now, indeed, it opens it again, this
scentless velvet,
crumbler-of-the-rocks,

this Lichen!

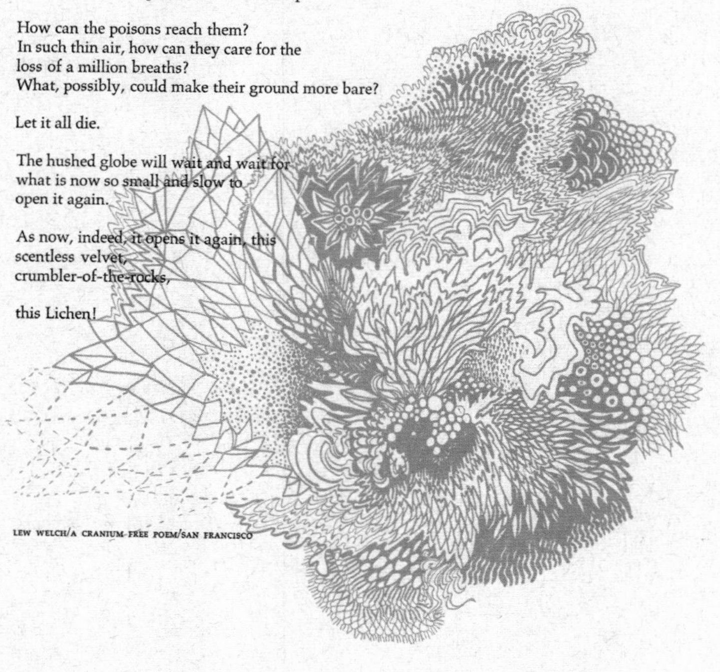

LEW WELCH/A CRANIUM-FREE POEM/SAN FRANCISCO

This broadside of Beat poet Lew Welch's poem was published in 1971, the same year that he disappeared in the Sierra Nevada, western United States, a presumed suicide.

Wolf lichen (*Letharia vulpina*) was used historically to poison wolves but was widely used as a dye and as a poultice for swellings and bruises.

The wolf lichen appeared as the title of an essay in Richard J. Nevle and Steven Nightingale's *The Paradise Notebooks: 90 Miles across the Sierra Nevada* (2022), the collaboration of Nevle, deputy director of the Earth Systems Program at Stanford University, and Nightingale, a poet, novelist and nonfiction writer. Their book is 'combinatory' (symbiotic) as they use essay and poetry to celebrate the Sierra Nevada, 'where gardens of black, orange, and chartreuse lichen adorn the rock' of the highest peaks. The wolf lichen, 'intensely chartreuse', whose vulpinic acid can both kill and heal, also absorbs contaminants from the air. Nevle sees in wolf lichen an 'injunction' to 'know how you are composed.' He calls this 'the practical and mystical expertise of the wolf lichen'. In gaining a solid understanding of the biology of the wolf lichen and combining it with our leanings towards spirituality, we may gain a similar expertise.[50]

Gander's colleague Brenda Hillman's collection *Extra Hidden Life, among the Days* (2018) celebrates life in the cracks of the Earth's surface.[51] The section 'Metaphor and Simile' has thumbnail photos of lichens in colour on almost every page: 'Some people think lichen looks dead but it is alive in its/ dismantling/ Some call it moss/ It doesn't matter what you call it./ Anything so radical & ordinary

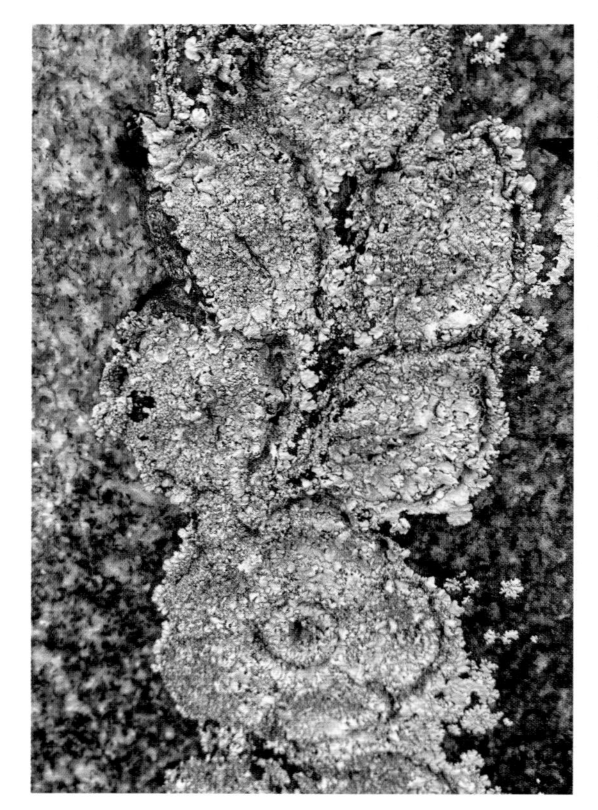

Lichens colonize etched features on polished granite gravestones and are said to protect porous limestone gravestones against water damage, Aurora, New York.

stands for something.'[52] Combining the words 'radical' and 'ordinary' describes lichen biology perfectly. Throughout the collection she references both common names and scientific names of lichens, and names the chemical tests used to identify them: 'cortex K^+ yellow, medulla K^-, KC^+ red to orange'.[53] Lichens model her activism. She writes, 'Lichen says accept what is then break it down.'[54]

We tend to notice mosses and lichens in cemeteries. So do Jack and Della, characters in Marilynne Robinson's lovely novel *Jack* (2020). During a long, enchanted night spent locked in a graveyard, they carry on a memorable conversation. Jack takes Della on a special detour in their wanderings to see his favourite cherub. Della

A damaged statue in the gardens of Birr Castle, County Offaly, Ireland, mended in green (moss) and orange (lichen).

says, 'I've seen prettier babies,' and he replies: 'There was moss on her lip a few weeks ago. It enhanced her metaphorical value, but it looked – uncomfortable. I used a toothbrush I brought here with me to clean her up a little.' They discuss whether 'terrified' or 'startled' is a better word to describe the cherub's expression. Della suggests that he replace the moss.[55] In a love poem called 'The Shampoo' Elizabeth Bishop (1911–1979) devotes the entire first stanza to lichens before addressing the act of shampooing. She describes them as 'still explosions on the rocks' that spread 'gray, concentric shocks'. Evoking lichens as other-worldly timekeepers conveys Bishop's feelings about the timelessness of her love.

Black lichens were an invitation to dark thoughts for Irish author Tim Robinson. When he walks the circumference of Árainn, one of the Aran Islands off the west coast of Ireland, he encounters 'a difficult mile'. The shore is 'surly', and he finds endless ankle-twisting contradictions underfoot:

> If then one could elevate gloom to metaphysical despair, see the human race as no taller than the most depressing of life-forms, the lichen that stains so many of these bare stones black, one might, paradoxically, march on with a weightier stride that would soon outwalk the linear desert.[56]

Robinson's inner life does not resonate with the black lichen during this difficult walk, but he continues, like a pilgrim, and finds that his 'perceptions are reborn'. Truly, a soft-bodied human seems out of place in this landscape of granite boulders and razor-sharp shards of broken rock, but not so for crustose lichens. The black wart lichen (*Verrucaria nigrescens*) is thought to hasten rock disintegration by elevating the temperature of the rock beneath it but also to protect rock surfaces from erosion by the ocean.[57] Lydia Davis's book *Our Strangers* (2023) includes a piece titled 'A Person Asked Me about Lichens', which begins, 'A friend asked me, had I ever written about lichens?' The answer is no, although she had taken notice of moss, seeds and pollen. By the end of her discursive thoughts on the subject she has become 'a lichen-curious person'. A lichen-curious person learns that lichens make good winter friends.[58]

In 'Patience', Chapter One of *H Is for Hawk* (2014), as Helen Macdonald confronts the death of her father, she remembers a time when she was nine. She is with her father, impatient to see hawks, and he explains the unwelcome need for patience:

> I got up slowly . . . and found I was holding a small clump of reindeer moss in one hand, a little piece of that branching, pale green-grey lichen that can survive just about anything the world throws at it. It is patience made manifest. Keep reindeer moss in the dark, freeze it, dry it to a crisp, it won't die. Impressive stuff. I weighed the little twiggy sphere in my hand. Hardly there at all.[59]

She keeps the piece of lichen by the phone as a memento of their outing. When she receives the call about her father's death, her gaze falls upon it. Poet Mary Oliver (1935–2019) finds 'spiritual patience' in observing moss in her poem 'Landscape' and calls moss her 'sweet cousin' as she touches it in 'The Moss'. Nikita Arora in 'The Many Meanings of Moss' (2022) writes that 'moss is touch'. Her essay answers in part Lew Welch's question, 'Why am I made to kneel and peer at Tiny?' She writes about touching moss on walks in Oxford:

> I happened upon a moss-covered log that glistened green under the overcast sky. The moss's leaves were as tiny and intricate as the finest embroidery, and as thin as clingfilm. I brushed my fingertips over the feathery bed in awe of its minuteness and complexity . . . Over the course of that winter, I touched mosses everywhere in the city: on footpaths and walls, on the barks of willows, on metal-based drain covers, on tombstones, on the roofs of houseboats, on abandoned bicycles, under the railway bridge.[60]

Touching moss takes her into the early days of moss research in Oxford, accounts of William Sherard and Dillenius and the origins of moss millions of years ago. 'This, then, is the first lesson that moss taught me: you can touch time.'[61]

While lichens do not invite touch as much as mosses – though they are wonderfully light and elastic when hydrated – they convey lessons as well. Bill Bryson, a writer who teaches us with humour and wit, includes lichens in his *A Short History of Nearly Everything* (2003):

> It is easy to overlook this thought that life just is. As humans we are inclined to feel that life must have a point. We have plans and aspirations and desires. We want to take constant advantage of all the intoxicating existence we've been endowed with. But what's life to a lichen? Yet its impulse to exist, to be, is every bit as strong as ours – arguably even stronger. If I were told that I had to spend decades being a furry growth on a rock in the woods, I believe I would lose the will to go on. Lichens don't. Like virtually all living things, they will suffer any hardship, endure any insult, for a moment's additional existence. Life, in short, just wants to be.[62]

Bryson thus finds sobering encouragement in his contemplation of the lichen. In *Ness* (2018), a modernist fable written by Robert Macfarlane and illustrated by Stanley Donwood, there is a mysterious character called 'she': 'her skin is lichen & her flesh is moss & her bones are fungi, she breathes in spores & she moves by hyphae. She is a rock-breaker, a tree-speaker, a place-shaper, a world-maker.'[63] Is she, this creature of moss and lichen, the spirit of Mother Earth? Perhaps Thoreau, one of Mother Earth's most attentive observers, deserves the last word. On 5 February 1853, as he entered Walden, he wrote: 'A thick fog . . . Not a bit of rotten wood lies on the dead leaves, but it is covered with fresh, green cup lichens . . . All the world seems a great lichen and to grow like one today . . .'[64]

seven

Curious Observers: A Field Trip

There is nothing, Sir, too little for so little a creature as man. It is by studying little things that we attain the great knowledge of having as little misery and as much happiness as possible.

SAMUEL JOHNSON, *Boswell's London Journals* (16 July 1763)[1]

There are certain sections of the public upon whose attention in particular we would strongly urge the claims of Lichenology or kindred studies . . . the invalid from our large towns, whose delicate mental and physical organization have suffered wreck in the too eager or engrossing pursuit of wealth or fame . . .

WILLIAM LAUDER LINDSAY,
A Popular History of British Lichens (1856)[2]

The great wordsmith and lexicographer Samuel Johnson may never have been attentive to a moss or lichen, but his point is pertinent to the engaged lifestyles achieved by bryologists and lichenologists. Likewise, while Japan's obsession with moss has been linked to the frenzied pace of modern life, there have always been men and women who have narrowed their focus to these small cryptogams, not just for a stroll but for a lifetime.[3] Few retire. Their personal stories now reside in the glowing obituaries of bryological and lichenological journals.[4] Some started young, some did not. Some were healthy, some battled afflictions as they worked. Some were academics, others were amateurs. Still others, in the spirit of citizen science, join the effort to record biodiversity before it disappears. Once attention is riveted, questions emerge, which lead to a cascading supply of further questions. Their stories connect us with stories of charismatic lichens and mosses and with the notion of biophilia, an idea conceived by Erich Fromm (1900–1980) and reintroduced by E. O. Wilson (1929–2021) as an innate urge to connect with other species. A field trip through the lives of some of these curious observers, whose gaze was held by mosses and lichens, leads all over the world.

William Starling Sullivant

William Starling Sullivant (1803–1873), a prominent businessman and passionate bryologist, was informally named by Harvard botanist Asa Gray as the 'Father of American Bryology' at the end of the nineteenth century. His father dying soon after his graduation from Yale, Sullivant returned to Ohio to take up the family business, working as a surveyor and engineer, managing a limestone quarry, grist mills, a sawmill and eventually a bank. At age thirty he started cataloguing the plants around his house, said to be inspired by a younger brother whose interest in natural history included 'botany, conchology, and ornithology'. Gray writes, 'As soon as the flowering plants of his district had ceased to afford him novelty, he turned to the mosses, in which he found abundant scientific occupation, of a kind well suited to his bent for patient and close observation, scrupulous accuracy, and nice discrimination.'[5] In 1843 Sullivant travelled the Allegheny Mountains from Maryland to Georgia with Gray, searching for mosses and producing *Musci Alleghaniensis* with his own copperplate engravings.

He found a botanical soulmate in his second wife, Eliza Griscom Wheeler (1817–1850).[6] Before her death from cholera at 33, she bore six children while working alongside him in all his botanical efforts. She shared botanical explorations, mounted specimens, wrote Latin descriptions and drew portraits of mosses like her husband, the eye having informed the hand as for Dillenius, Hedwig and Hofmeister. William Schimper, one of the foremost authorities on

Alongside civic and business responsibilities, William Starling Sullivant travelled the Allegheny Mountains in search of mosses, establishing the foundations of American bryology.

mosses in Europe, admired her drawings so much that he named a moss after her. Sullivant's close collaborator in Ohio, Swiss-born Leo Lesquereux, would write of him after death,

> In everything, as you well know, W.S.S. was most accurate. He was superficial in nothing. He worked his mosses slowly, coming again and again to a doubtful species, comparing authorities, repeating the most difficult anatomical preparations, til fully satisfied that his conclusions were warranted as far as botanical science could warrant them.[7]

Before arriving in the United States, Lesquereux had studied peat bogs throughout Europe. The Sullivant Moss Society, named in his honour, eventually became today's American Bryological and Lichenological Society.

William Borrer

William Borrer (1781–1862), known as the 'Father of British Lichenology', knew and appreciated higher plants of all kinds, but chose to specialize in lichens. He settled at Barrow Hill in Sussex, where he planted a garden with 6,600 species. Bryological historian Mark R. D. Seaward writes, 'Borrer had a particular flair for notoriously difficult groups and was renowned for his knowledge of *Rosa*, *Rubus*, and *Salix*, genera which to this day taxonomically perplex all but a few.'[8] But it was through the lichens that he could truly put his flair to use. He gained a name throughout Europe, even with Acharius, 'Father of Lichenology', in Sweden. He corresponded extensively, sharing findings and confirming identifications. He is known as a field

William Borrer studied enigmatic dust lichens (*Lepraria* sp.) such as these, which colonize old bark, rock and shaded embankments and apparently never develop reproductive structures (apothecia).

botanist who detected lichens never before noticed, one who spared neither time nor expense in verifying a sighting. If he saw a specimen from the train, he would hop off at the next stop and walk back to confirm his observation. *Lichenographia Britannica* (1809–39), a project he began with Dawson Turner, a wealthy banker in Yarmouth who was intensely interested in cryptogams, remained incomplete, largely due to Turner's busy life. However, Borrer's descriptions of lichens in words are compelling 'portraits' that impress modern-day lichenologists who now use the tools of advanced microscopy, colour tests and chemistry to identify lichens. A description of the crustose lichen *Lepraria viridis* shows that he noticed that some lichens failed to thrive in smoky towns 56 years before William Nylander, who is generally given credit for first observing the connection between air pollution and lichens in his studies of lichens in Luxembourg Gardens.[9]

An anonymous 'In Memoriam' eulogy attests copiously to 'the truth-loving character of all his communications'.[10] The eulogist seems to particularly relish the job of praising a botanist: 'The votaries of Flora are not much tormented with the spleen, the megrims, and other sullen vapors which becloud the brain and enchain the spirit. They inhale the invigorating breezes of the hills, the freshness of the fields teeming with vegetation, enjoy the ever-changing aspects of nature.'[11] Acharius described him as 'Lichenologus eximius', *eximius* meaning extraordinary. Perhaps his extraordinary gifts related to intuitive knowledge gained from extensive fieldwork.

William Mitten

Sussex was fertile ground for students of mosses as well as lichens. William Mitten (1819–1906) was born just 8 kilometres from William Borrer in Hurstpierpoint, Sussex. He apprenticed to a chemist in Lewes before returning to Hurstpierpoint, where he became a dispensing chemist, working long hours (7 a.m. to 10 p.m.) for 56 years. During his apprenticeship at Lewes he annoyed his employer by stuffing botanical specimens into the pigeonholes reserved for various formulations.[12] Borrer became his mentor in every way, lending him a microscope, giving him the run of his library and herbarium and sharing his ways of looking at plants. When Mitten was 27, he discovered near Hurstpierpoint the rare moss *Weissia (Astomum) mittenii*, which was immediately added to the *Bryologia Europea*. It has not been found in Britain since 1920 and is considered a victim of agricultural encroachment.[13] *Weissia* species favour bare ground and can even form moss balls in certain circumstances. Mitten would say, 'To Mr. Borrer I owe the ability to determine with exactness.'[14]

While Borrer botanized extensively in the UK, Mitten would become known as the bryologist who travelled the world looking at mosses while never leaving Europe.[15] Borrer had introduced Mitten to William and Joseph Hooker of Kew as a rising star in bryology,

By all accounts William Mitten lived the life of a hard-working pharmacist, a well-loved community member and an eminent bryologist with grace.

but Mitten turned down a job at Kew as a curator in order to care for his wife and four young daughters, pharmacy being more lucrative than herbarium work. However, Kew began forwarding the mosses received from colonial collectors abroad to Mitten for determination, a practice that continued for fifty years. One of them was the superintendent of the botanical gardens in Peradeniya, Ceylon, George Thwaites, who had made such pertinent observations about lichen symbiosis. In 1851 Mitten published his first determinations of mosses from Quito, Ecuador, forwarded to Kew by William Jameson, a Scottish physician and botanist living abroad. This was followed over the years by mosses from around the world – Portugal, India, Ceylon, Tasmania, New Zealand and South America, for example – and even several collections from the volcanic Kerguelen (Desolation) Islands made during the visit of HMS *Challenger* during its grand tour of 1873–6 and the Transit of Venus Expedition of 1874. Mitten also enjoyed occasional field trips in Sussex. Vivienne Manchester, historian of Hurst, describes the appearance of Mitten and his good friend Bishop James Hannington ('later martyred in Uganda') after a field trip along the Gotthard Pass to Hospental, Switzerland,

> where they crawled on their hands and knees, with soil-stained pockets bulging with leaves, stalks and trailings of moss. They were dressed in old suits, baggy at the knees and elbows, and James had on his head a grass hat, like an inverted flower pot, with a pocket hanky wound loosely around it to shade his sun-scorched face.[16]

Specimens awaiting determination piled up in the pharmacy in what British Bryological Society historian Brad Scott calls a 'tsunami'. Scott notes that 'His painstaking work illustrates the vital part specialist local practitioners played in the Hookers' imperial botanical project.'[17] By the end of his life Mitten had amassed a personal herbarium of 50,000 specimens of mosses, lichens and liverworts (he always received duplicates for comparison with future moss deliveries). Just

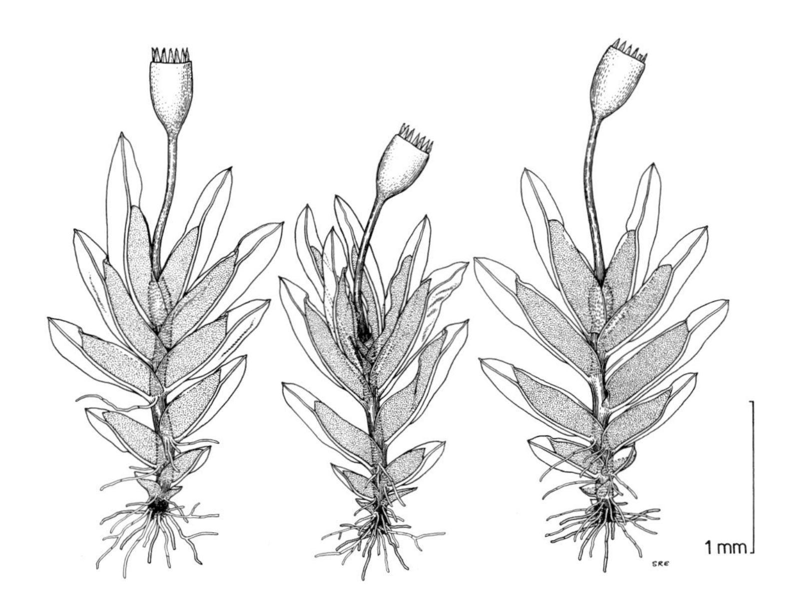

Fissidens parkii was named by William Mitten after explorer Mungo Park, who declared himself saved from death by noticing 'the extraordinary beauty of a tiny moss', which he collected 'mid-day, 25 August 1796' in Mali; drawn by bryologist Sean Edwards.

days after his death the New York Botanical Garden acquired his herbarium and his 'exquisite drawings' for £400.[18] Although Kew's Joseph Hooker had written to Mitten in 1885 angling for custodianship, in 1891 Elizabeth Gertrude Knight Britton of the New York Botanical Garden personally visited Mitten, securing his agreement.

As Mitten gained acquaintance with the mosses of the world, he placed greater emphasis on leaf characteristics than on those of the peristome for classification. He gained a reputation for being more of a splitter (one who names new species) than a lumper (one who merges species), but many agreed that this was to be expected at a time when so many new mosses were being discovered, and that he was 'gifted with an unusually critical faculty in discriminating between closely allied species'.[19] Even today bryologists query whether splitters simply see characteristics that others cannot. Mitten was fortunate to travel the world of mosses while living at his family home, Treeps, with his wife and daughters, where he gathered lichens from the graveyard next door, an interest he had picked up from Borrer.[20]

Ellen Hutchins

Many talented women have fallen for cryptogams, one being Ellen Hutchins (1785–1815), now known as Ireland's first female botanist – a woman of enquiring mind and discerning eye who made significant discoveries about cryptogams.[21] Hutchins lived heroically, threading the needle between the work of family and the work of living in the world as naturalist, plant hunter and botanist. One of 21 children, she was born near the Bay of Bantry, a remote and wild area of West Cork surrounded by mountains. Her father, a magistrate, died intestate when Hutchins was two, leaving the family in financial disarray, torn apart by conflict between the elder brothers. Hutchins spent a brief period at a boarding school in Dublin but returned home because of illness. Dr Whitley Stokes (1763–1845), who treated her in Dublin, was an enthusiastic amateur botanist who advised her to undertake study of a branch of natural history to ease isolation and improve her health. To this end he lent her botany books. At that time the flora of West Cork was unknown and unnamed.

She began studying and collecting cryptogams – seaweeds, mosses, lichens and liverworts – in her spare time. A partial invalid herself, she became responsible for the care of her invalid mother and brother Thomas, who had to be carried everywhere. Her correspondence makes clear that this was exhausting work, night and day. However, she educated herself. Soon she had books, a microscope, a hammer for dislodging 'morsels' of lichen from rocks and even access to a boat and crew for finding seaweeds. She sent specimens to Dr Stokes, who sent them on to James Mackay at Trinity College Dublin and subscribers such as Dawson Turner. It was the practice of the day for botanists to exchange specimens and to share insights about their names and affinities in letters. She gained a reputation in wider botanical circles for her collections and exquisite botanical drawings. Soon 'Mr Mackay' came to visit her. Her knowledge extended to higher plants as well. As requested by her mentors, she compiled a list of 1,100 plants, the first flora of that area of southwest Ireland.

James Edward Smith, founder of the Linnean Society, said that Hutchins could find almost anything and William Jackson Hooker, named director of Kew in 1841, wrote that 'Miss Hutchins is a mine, but I never intend to bore her lest she should be too prolific.'[22] She had contributed so many specimens for his monograph on a family of leafy liverworts (the British Jungermanniae) that he told Turner: 'Miss Hutchins' discoveries alone will form an Appendix as large as the work itself.'[23] Bantry Bay figured as habitat for half of the leafy liverworts in his monograph. Overwhelmed by her prodigious fieldwork, Hooker praised her in the published volume and named a beautiful liverwort after her.

Though she loved searching for seaweeds in the bay, she looked forward to finding lichens on land: 'I hope to visit the mountains. I expect to find many Lichens, I never go out without finding something new.'[24] Turner would send her lichens to William Borrer, for his opinions, which Turner would pass on to her with his disagreements with Borrer's determinations noted in the margin. On the last page of his monograph on seaweeds, *Plantarum Fucorum* (1808–19), Turner acknowledges her as his 'able assistant'. He writes that should he ever finish *Lichenographia Britannica*, which he had worked on for so long with Borrer, he would surely acknowledge her there as well, because 'Botany had lost a votary as indefatigable as she was acute, and as successful as she was indefatigable.'[25] He never finished, perhaps because he lacked his able assistant. In her last letter to Turner, when she was very ill, she wrote 'send me a moss . . . anything to look at.'[26]

An interchange between Hutchins and Turner about a specimen reveals the difficulty of 'determining' or naming lichens. In a letter of 8 March 1811 she writes, 'The Opegraphae are very puzzling. I think many of my No^{s}. are I am sure only varieties, but I grope in the dark,

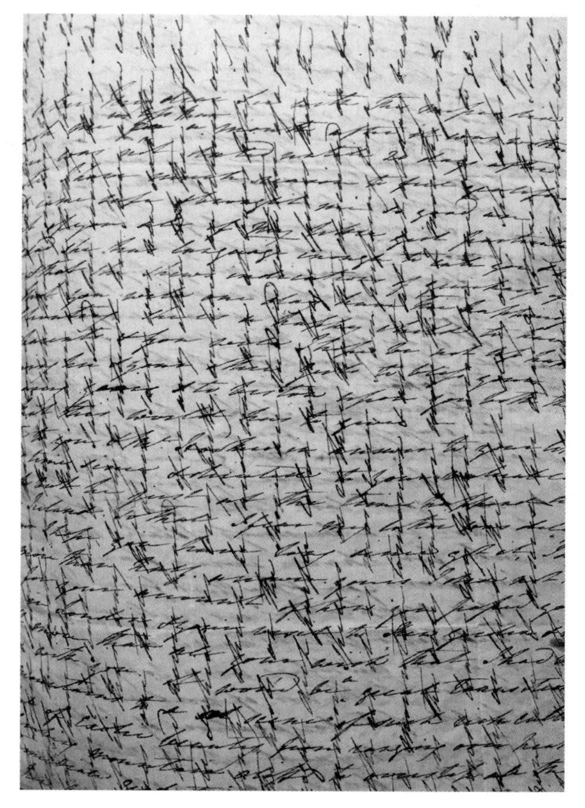

One of Ellen Hutchins's letters to Dawson Turner, cross-hatched to save postal charges for the recipient, in which she writes that should she ever go to Kew Gardens, it might be too 'overpowering' to encounter such 'Exotic beauty' after so long 'ranging over heath & bog'.

& can only distinguish by my own observations without descriptions to guide me.'[27] He replied in a letter of 25 March 1811 that:

> The Opegraphae are indeed puzzling: I once amused myself with writing a monograph of them, & thought I had succeeded à merveille, but when after a short respite I tried to examine some of the species by my own characters, there was so much room for discontent that I have never opened the papers since.[28]

Members of the genus *Opegrapha* are called the scribble lichens. They are crustose lichens distinguished by dark elongated marks, scribbles (lirellae), which are the fungal partner's spore-producing structures (apothecia). A new species, *O. halophila,* from coastal British Columbia, Canada and Alaska, was named in 2019. It produces the distinctive roccellic/angardianic acid.[29]

Richard Spruce

The person whose moss collections inundated Mitten most thoroughly was Richard Spruce (1817–1893), equally passionate about mosses but quite unlike him in temperament and life experience. Spruce spent fifteen years collecting plants in South America under contract from Kew and the East India Trading Company, travelling

Though brilliant in linguistics and mathematics, Richard Spruce led the life of a plant explorer, discovering new mosses in both South America and Britain.

alongside Alfred Russel Wallace, co-founder with Darwin of the theory of evolution, and Henry Walter Bates (1825–1892), author of *The Naturalist on the River Amazons* (1863). Spruce made a name for himself as a young man by compiling a list of the mosses near and around Ganthorpe, Yorkshire, where he was born. He followed the practice of making extensive lists of plants wherever he went. This brought him to the attention of Kew, who needed collectors. He was an unlikely plant hunter, being frail throughout life, enfeebled by infections and fevers and racked by a constant, bloody cough. Elaine Ayers describes Spruce as a person who never achieved the fame he deserved because mosses were not compelling for the Victorians; they were more impressed by the brightly coloured orchids and other flowers brought back from the tropics.[30]

Though he achieved some renown as the collector of cinchona, which was transported to India as a cure for malaria, it was his personal connection with mosses and liverworts that sustained him during the travails of his health and the discomforts of botanical travel. He wrote, 'Throughout the journey, whenever rains, swollen streams and grumbling Indians combined to overwhelm me with chagrin, I found reason to thank heaven which had enabled me to forget for the moment all my troubles in the contemplation of a simple moss.'[31] Ayers suggests that by the end of his life Spruce shows himself 'contemplating his own gradual transformation into the object of his fervent study: moss'. She quotes a passage from Spruce's letter to a friend: 'One day last week a dentist relieved me of four teeth, and I now belong to the genus *Gymnostomum*; but by the time you come over I hope to have developed a complete double peristome.'[32] This was the publication year of Mitten's major work, the 632-page *Musci Austro-Americani* (1869), based mostly on Spruce's mosses from South America. Spruce expressed his biophilia in this comment:

> I like to look on plants as sentient beings, which live and enjoy their lives – which beautify the earth during life, and

> after death may adorn my herbarium . . . It is true that the Hepaticae have hardly as yet yielded any substance to man capable of stupefying him, or of forcing his stomach to empty its contents, nor are they good for food; but if man cannot torture them to his uses or abuses, they are infinitely useful where God has placed them, as I hope to live to show; and they are, at the least, useful to, and beautiful in, themselves – surely the primary motive for every individual existence.[33]

Naming as a way of signalling connection, as well as their passion for beauty, drove Spruce and Mitten to extraordinary labour.

Elizabeth Gertrude Knight Britton

Elizabeth Gertrude Knight (1858–1934) was born in New York City on 9 January 1858 to a family that owned a furniture factory and a sugar plantation in Cuba. Her childhood in Cuba fostered a love of plant life. Joining the Torrey Botanical Club as a young woman revealed exceptional talent for fieldwork. Finding the 'fruit' (capsules) of the rare moss *Eustichium norvegicum*, which had eluded Sullivant, in the 'dells' of the Wisconsin River on 8 July 1885, determined her future. She became the leading bryologist of her day, though much of her work was unofficial, unpaid and honorary, like that of Annie Lorrain Smith. When she married Nathaniel Lord Britton, a geology professor at Columbia College, she took an unpaid position at Columbia as moss collector. During her life she would travel throughout the United States, the Bahamas, Puerto Rico, Cuba and Europe. She completed *A Catalogue of the Mosses of West Virginia* in 1892 and a series of articles, 'How to Study the Mosses', in 1894. She was a popularizer and showed endless energy for establishing organizations that would serve the public and then serving them herself as fundraiser, secretary or president. She and her husband spearheaded the founding of the New York Botanical Garden (NYBG) after their visit to Kew. She published over 340 papers and had doctoral students, even though

Elizabeth Gertrude Knight Britton established a formidable reputation as an outstanding bryologist, despite having no formal degrees.

she had no advanced degrees. In 1898 she and her student Abel Joel Grout founded the Sullivant Moss Society, which became the American Bryological and Lichenological Society. She might have been inspired by the Moss Exchange Club, now the British Bryological Society, begun two years earlier in England for the purpose of helping members correctly name specimens in their collections. The moss genus *Bryobrittonia* is named after her.[34] After her death a society member would write, 'The beginnings of the limited popular interest in mosses which led to the formation of our Society must be attributed directly to that dominating, or even magnetic personality, Elizabeth Gertrude Knight Britton, whose name was on every bryologist's tongue for 40 years.'[35]

One of her many legacies lies in the work of Abel Joel Grout, who would receive a doctoral degree under her tutelage and wrote such classic books as *Mosses with Hand-Lens and Microscope* (1903), *Mosses with a Hand-Lens* (1905) and *The Moss Flora of New York City and Vicinity* (1916). After retirement from teaching he received visitors at Moss Rock Cabin in Newfane, Vermont, where bryologist Seville Flowers, visiting from Utah, had memories of 'rock walls, lichens, a crooked rain pipe, seeing *Buxbaumia* growing for the first time, Baker's brook, paper birches, lycopods, mosses'.[36]

Matilda Cullen Knowles

Some fifty years after Ellen Hutchins's death Ireland produced another notable lichenologist, Matilda Cullen Knowles (1864–1933), who would both 'determine' and 'publish'. Her major work was *The Lichens of Ireland* (1929), an effort aided by thirty collaborators, who had by this time proliferated, and included the distributions of eight hundred lichens in Ireland. She is especially known for a detailed study of the ecology of lichens on the sea-cliffs of Howth Head, a peninsula forming the northern boundary of Dublin Bay, Ireland.[37] She describes the zones or 'belts' of lichens occurring at increasing distance from the shore that were distinguishable by colour – black, orange and grey. The black lichens (*Verrucaria* sp.) occur at shoreline, while the widest belt was populated by numerous species of grey-green *Ramalina*, some known as just A, B and C because they defied determination at the time. They are fruticose macrolichens and have tubular 'stems' that are flattened. In her paper she refers to herself as 'the authoress'. The 'authoress' wonders why *Ramalina* B predominates in conditions that are 'inhospitable' even to the undemanding crustose species. In the bright sun of spring and summer it becomes 'burnt up and brittle'. Looking even more closely, she found a 'subvegetation' of crustose and foliose forms hiding in the shelter of the *Ramalina* fronds. Ireland's National Botanic Gardens honoured Knowles's many years of work in its herbarium with a commemorative plaque in 2014.[38]

A note about *Ramalina*: The ramalinas are commonly known as strap or cartilage lichens. One of the most beautiful is the lace lichen

The net-like growth pattern of lace lichen (*Ramalina menziesii*) is remarkable even for a lichen.

of North America, an important forage for deer in the Coast Range of California. California named it the State Lichen in 2015, the first state to honour a lichen. The authors of *Lichens of North America* write that 'some mule deer have been known to fight over it.'[39] A study of herbarium specimens revealed lead contamination, which peaked in the 1970s owing to leaded petrol and has since decreased. However, specimens are now showing high concentrations of monomethylmercury (MMHg) absorbed from the coastal fogs and transmitted to the deer and their predators (puma) through a process biologists call biomagnification. Kawaiisu, an indigenous tribe of south-central California at the southern end of the Sierra Nevada, called it *paazíiomoora*. It was used by rain shamans, who considered it unpredictable, bringing unwelcome kinds of precipitation, and so powerful that it might put them out of business. Mothers hid it from children, fearing the results of playful rain-making. Its use in this regard was recorded as late as the 1970s.[40]

Elke Mackenzie

'The Unsung Heroine of Lichenology' is the title of science writer Sabrina Imbler's 2020 profile of the remarkable Elke Mackenzie (1911–1990). Born in London in 1911 and educated at Edinburgh University, Mackenzie apprenticed with lichenologist Annie Lorrain Smith at the Natural History Museum in London, and later became director of the Farlow Herbarium at Harvard until retirement. A trip to Antarctica with Operation Tabarin during the Second World War cemented her interest in lichens. Here she prepared 865 specimens to bring home for study and examined mosses and lichens, especially

Snow, or foam, lichens are among the most architecturally complicated lichens; Easter lichen (*Stereocaulon paschale*) on a sandy roadside in Jämtland County, Sweden.

admiring 'vivid green' cushions of the moss *Polytrichum*, 60 centimetres across, but the love of her life would be members of the lichen genus *Stereocaulon*.[41] Variously called snow lichens, foam lichens or Easter lichens, they are hard to describe. Lichenologists use some unusual adjectives to explain their morphology: caespitose (growing in dense tufts), verrucose (covered in warts) and subbotryose (clustered like grapes). They create vast mats of gnarly, greyish-white mounds in boreal landscapes. One species, *S. arenarium*, grows on moss from Greenland to Kamkatchka. No doubt the intrepid Mackenzie would have loved seeing *S. ramulosum* cozily sharing an earth bank with moss in Australia's lichen-filled Wombat State Forest.[42]

Rebecca Yahr

While 'determining' lichens is ongoing, much of the work of the lichenologist in the twenty-first century concerns public education and conservation. Rebecca Yahr at the Royal Botanic Garden Edinburgh holds the title of Lichen Biodiversity Specialist. She teaches lichen courses for the general public and leads lichen safaris in the gardens, hoping to inspire interest in lichen diversity:

> Working on a specialist group like lichens, I feel called to help people discover the diversity of the world around them – indeed probably on their windowsills and pavements – not just in the more famous ecosystems such as rainforests. With a good teacher and a bit of interest, the diversity of life on earth is forever fascinating, and I really enjoy helping people find that fascination.[43]

In 2022 Yahr became the subject of a profile in *The Gentlewoman*, a magazine that usually features fashion designers, writers, actresses and singers, an indication that the public is ready to be fascinated with lichens.[44] In 2011 she and a colleague named a new species in *Opegrapha*, the genus that Hutchins and Turner found so mystifying.

A slender twig provides substrate for a diverse group of lichens, Highland County, Virginia.

The new species, *O. viridipruinosa*, was found on the bark of an elderberry growing in an abandoned chalk pit in East Suffolk. Yahr posted a photograph of a lichen-encrusted twig on X accompanied by the following commentary on 7 January 2023: 'Why do I like lichens? For the stories they tell. This run of the mill branch has some ordinary lichens doing some fascinating things.' She then wrote her own thread, a lesson in lichen ecology, pointing out that a black line between two different crustose species represents 'a zone of stable interaction', while a leafy lichen showing bleaching indicates a crustose species making a defensive move and a hint of moisture around a lichen on dry bark shows how lichens trap water trickling down a trunk.[45]

The Sussex Bryophytes

Since the days of Borrer and Mitten, Sussex has continued to be fertile ground for those interested in mosses and lichens. While local groups throughout the world go out on moss and lichen forays and rambles, the Sussex Bryophytes, a chapter of the British Bryological Society South East, has a particular census mission. Their goal is to record forty bryophytes in every 2 × 2 kilometre tetrad of each 10-kilometre square in East Sussex. There were over three hundred tetrads in East Sussex with very few records when the group set to work in 2016. The group has averaged 2,000 new records every six months. Blogs describing the routes taken and names of species encountered capture the spirit of adventure animating these volunteers. Sometimes the writers mention the mosses they know like old friends. For instance, after a visit to Bexhill Down, Sue Rubenstein writes,

> Between a school and the park were some areas of bare, scraped earth. Lots of little capsules of pale green *Tortula truncata* and one small area of brown *Microbryum davallianum* filled the area. Lime green patches were *Pseudocrossidium hornschuchianum* while *Bryum ruderale* had dark purple rhizoids and smooth, pale red tubers.[46]

The moss *Pseudocrossidium hornschuchianum* is an old friend of Sussex Bryophytes recorders.

David Newman went to tetrad TQ63Y near Scotney Castle Estate in January 2023 on a windy day and noted that 'exploring a new area and seeking out bryophytes always enlivens the stuttering grey cells'. On this occasion he attracted the attention of two horses, one of whom bit him repeatedly and painfully, which brought the outing to an end.[47] Usually the field trips are restorative, however, as member Brad Scott writes in a blog post titled 'Zen and the Art of Bryological Recording':

> Bryological recording walks come in many guises. They may be solitary, in pairs, small groups, large groups, in the rain, or the sun, or even the snow. Sometimes you wonder quite why you are there, and on other occasions they can be the most wonderful moments, either because of finding an interesting species, or place you did not expect, or simply because of the sublime experience of being in the present moment and meeting small plants.[48]

Scott has studied the history of bryological recording going back to the collections of Dillenius in the early eighteenth century.[49] As he explored the farmland and woods around Perryhill, he too met a few old friends ('common species of gravelly paths and soils'): '*Streblotrichum convolutum* . . . turned up almost immediately, growing with its even brighter green neighbour *Pseudocrossidium hornschuchianum*' and completed 'a gentle 3½ mile mosey along the lanes'. He writes at the end of his account that he had begun his field trip with a thousand different thoughts in his head, but they dissipated as the outing proceeded: 'there wasn't even much of a striving for bryophytes, either; I walked, I looked, and there they were.'[50]

Kerry Knudsen

Kerry Knudsen, curator of lichens at the University of California Riverside and mycological researcher at the Czech University of Life Sciences, says that he, like Scott, becomes 'completely visual' and 'non-rational' when looking for lichens and that looking at lichens is easier than looking at redwoods.[51] Disabled at 42 after twenty years in the construction industry, he turned to his back garden for an object of study while he stayed home as 'Mr Mom'. It was going to be either mosses or lichens, but he found a few lichens first. Now a world-renowned specialist in the (mostly) crustose genus *Acarospora*, cobblestone lichens, he and his lichenologist wife, Jana Kocourková, do extensive fieldwork in the Chihuahuan, Mojave and Sonoran deserts of North America. Described as an 'ex-anarchist' in another profile, he notes how he looks at nature as a surrealist.[52] He works six days out of every week studying lichen biodiversity, currently threatened by uncontrollable wildfires in North America, because their appearance delights him – they 'look like fairy tales', he says in Matthew Killip's filmic profile.[53]

The attitudes and work endeavours of bryologists and lichenologists suggest the mindset of biophiles. Kay Harel uses Darwin as an example to make the case that those who are curious are biophiles.

The bright cobblestone lichen (*Acarospora socialis*) is a pioneer species, occurring on soil crusts and rocks in the deserts of western North America.

In other words, there is a connection between curiosity and love of other organisms. She writes that 'It is axiomatic that only the curious see something new, and Darwin's curiosity was extravagant.'[54] She also reminds us that philosopher William James believed that we all have some of the qualities that make certain individuals extraordinary. Erich Fromm had seen biophilia as a complex of attitudes – truth-loving, adventurous, wondering, curious. Those individuals described in this chapter now 'living' in obituaries travelled the world fuelled by curiosity, their lives constant field trips into the unknown. They laboured over making accurate determinations in a quest for truth. In 'The Role of the Amateur in Bryology: Tales of an Amateur

Bryologist' (2003) Kenneth Kellman, a commercial air conditioning mechanic by trade, describes how he became a knowledgeable amateur bryologist despite having no formal training in botany and no degrees. He writes that he gained expertise from 'a persistent curiosity and the good will of more knowledgeable friends'. His closing remark is 'I encourage you to polish your hand lens, and then open your eyes to the tiny world that you have been walking by every day of your life.'[55] An amateur is, etymologically, one who loves. It's hard to go off the rails in a life spent outdoors, dedicated to peering at the near invisible on bended knee.

Iwatake (*Umbilicaria esculenta*), a rock tripe lichen used as tonic food in Asia for centuries, grows on rocky cliffs; Utagawa Hiroshige II, *Iwatake Gathering at Kumano in Kishū*, 1860, colour woodblock print.

eight
#moss#lichen

In conclusion, bryophytes, with the exception of peat mosses, are of limited value to man when they are considered from a purely utilitarian or monetary point of view. Perhaps their most important direct use is as experimental plants. However, when their role in the economy of nature is examined, they have a great, though often indirect, value to and influence upon man.
JOHN W. THIERET, 'Bryophytes as Economic Plants' (1956)[1]

Fifteen years ago I had an odd dream. In it, a medicinal plant that I was interested in, an Usnea lichen that is ubiquitous on trees throughout the world, told me that while it was good for healing human lungs it was primarily a medicine for the lungs of the planet, the trees. When I awoke, I was amazed. It had never occurred to me in quite that way that plants have some life and purpose outside their use to human beings.
STEPHEN HARROD BUHNER, *The Lost Language of Plants* (2002)[2]

Mosses and lichens have supported the growth of human culture in many activities: survival (famine food), brewing, baking, biomonitoring, caulking, decorating, embalming, feeding livestock, diapering, dyeing, healing (materia medica), packing, perfuming, powdering (hair), poisoning, sniffing, snuffing, spiritual/ritual practice, stuffing, tanning and so on. These uses reveal ingenuity, imagination and desperation: the Wylackie (Wailaki) people, who lived near Eel River, California, used the pitted beard lichen (*Usnea cavernosa*) wrapped around the brains of

Neckera douglasii, a western North American relative of *Neckera crispa*, the dominant species used for caulking in Europe from the early Bronze Age onwards.

a deceased animal to tan its hide; in the 1700s people in Småland, Sweden, collected *Lichen candelarius* (*Candelaria* sp.) to dye candles 'golden' for religious ceremonies; and in the early 1800s farmers in Virginia collected *Usnea* to feed their livestock.[3] Neither moss nor lichen grow fast enough to serve commercial purposes in the modern world. Perhaps we can view ourselves not as extractive users, but rather as symbiotic partners, like the birds mentioned below.

#medicinalmoss

The study of the plants used for medicinal purposes by aboriginal/native cultures was given the name 'ethnobotany' by American botanist John Harshberger (1869–1929) in 1896. Ethnobryology and ethnolichenology have become subsets of this field. Gathering information requires painstaking work by anthropologically trained botanists, who must learn little-known languages and visit remote areas to gain information accessed only through an oral tradition. Most reports relating to medicinal mosses come from western North America, Canada and China. The reviews made thus far document between 150 and 170 uses for over 100 taxa (sometimes including liverworts), with applications to most areas of the human body.[4]

Early users of medicinal remedies developed methods of classification. Northwestern North American cultures grouped and subdivided mosses according to habitat. The Tzeltal of Mexico and the Okanogan-Colville First Nations of Canada named bryophytes by their location – for example, ground, rocks, wood, trees, caves and water. The Oweekeno First Nation of British Columbia preferred *Plagiomnium insigne* growing under spruce rather than Douglas fir trees for its greater medicinal value. The early English herbalists also found medicinal significance in the type of tree upon which a moss or lichen grew, as noted in Nicholas Culpeper's *Herbal* of 1653.[5]

It is thought that mosses were used medicinally in China by the sixth century. The moss *Polytrichum* (TuMaZang, or Horse-Mane-of-the-Earth) was mentioned for use as a diuretic in the *Compendium of Materia Medica* (drafted 1578, printed 1596) written by Li Shizhen (1518–1593). The lovely moss *Rhodobryum gigantea* ('return-the-heart-herb') was used for heart healing. It is still used in southwestern China to treat cardiovascular problems.[6]

Researchers at the Medical University of Silesia in Poland recently took a deep dive into the past to uncover the identity of a moss recommended for use in wound healing in 1651. The passage in Jean (Johann) Bauhin's *Historia plantarum universalis*, translated from the Latin, reads: 'Practitioners use this moss . . . for stemming the blood, learned from bears, which, whenever hurt, use them to stop blood.'[7] Detective work revealed a plethora of aliases resulting from the introduction of ever lengthier polynomials by different observers, for example, *Muscus querno vilissimo vilior, saxis et udis terrae glebis adnascens*. Finally, they tested three suspects for absorption capabilities. One, *Brachythecium rutabulum*, the rough-stalked feather moss, tested much higher than the other two candidates, with 1 gram of air-dried moss absorbing just over 16 grams of water. Their literature search yielded data on its antifungal, antimicrobial and anti-human carcinoma cell-growth properties and its ongoing use in the Himalayas in an ointment for burns and wounds.

The star of medicinal mosses is 'the kindly sphagnum' of the First World War (1914–18). The story of how sphagnum moss saved lives among the 20 million wounded will never be forgotten.[8] Those at home on both sides of the Atlantic searched bogs for sphagnum, which they collected, sorted and stitched into surgical dressings that were sewn into the inside of soldiers' uniforms for quick use on the battlefield. Though the injured have stuffed sphagnum into wounds since the Bronze Age (3300–1200 BCE), it was in Edinburgh, in 1914, that surgeon Charles W. Cathcart organized efforts to use sphagnum moss in dressings to replace cotton wool from Egypt, which had supply chain issues and was also in demand militarily for gun cotton and nitrocellulose explosives. Cathcart campaigned vigorously for its absorbent capacities, advertising that 283.5 grams of dried moss could hold 1,984.5 grams of water. Sphagnum had also long been known to have antiseptic and deodorizing properties. Collaborating with his friend Isaac Bayley Balfour, keeper of the Royal Botanic Garden, Edinburgh, Cathcart won approval from the War Office to accept sphagnum dressings as 'Official Dressings'. By 1918 the British were making 1 million pads per month. Collection across the UK occurred from Bodmin Moor to Shetland, with women and children leading the way.

This soon became an international effort. When the French Relief contacted the University of Washington in Seattle in July 1917 for supplemental dressings, botanist J. W. Hotson became superintendent of moss dressings for the Northwest Division of the Red Cross. His detailed 1918 report titled 'Sphagnum as a Surgical Dressing' shows how carefully the biology of sphagnum was explored to make the most effective pads.[9] Making sphagnum dressings on such a scale was never feasible again for a number of reasons, no doubt among them because an army of volunteers to search bogs and hand-pick the gatherings was lacking and because no machine could handle the tiny, fragile leaves with the required care. It was not until the 1990s that British chemist Terence J. Painter's experimental studies verified the antimicrobial capacities of sphagnan, just one of

Waldport School children, teachers, board members and volunteers gather sphagnum moss for surgical dressings during the First World War in Warren Spruce Company's right of way, Portland, Oregon, 1917–18.

sphagnum's secondary compounds.[10] Now we know that sphagnan, a pectin-like component of the walls of sphagnum's hyaline cells, has the capacity to immobilize pathogens through a cross-linking reaction, rendering them inactive.[11]

#medicinallichen

Linnaeus was a professor of practical medicine as well as a professor of botany. In this capacity he published an influential medical textbook in 1749, in which he included seven lichens: *Lobaria pulmonaria* for coughs and jaundice; *Usnea* spp. for wounds, bleeding and umbilical hernia; *Cetraria islandica* for fever; *Cladonia pyxidata* for whooping cough; *Peltigera canina* for hydrophobia (rabies); *Peltigera aphthosa* for thrush and worms; and skull lichens for epilepsy and haemorrhages. His list was based on hundreds of years of use going back to Dioscorides and Galen. Linnaeus' continued endorsement meant that for the next one hundred years pharmacists and chemists throughout Europe and Scandinavia were grinding lichens into powders, boiling them into pastes, studying their mucilaginous residues and promoting their curative properties. The contemporary use of several of these lichens has been documented in eastern Andalucia, Spain.[12]

The doctrine of signatures, an ancient method of bioprospecting based on analogy, which was also used in Chinese medicine, inspired the early practices. The reasoning was that plants revealed through one or more characteristics which part of the body they could heal by resemblance to the body part, for example, the red colour of sap indicating a cure for blood ailments. The lichens on Linnaeus' list were constants in the herbals that began to proliferate in the late Middle Ages. In the seventeenth century Thomas Willis (1621–1675), an influential 'chemical physician', recommended a preparation that used the pixie cup lichen, *Cladonia pyxidata*, for the much-dreaded 'chin-cough' (whooping cough or pertussis). It contains, he wrote, 'particles somewhat sharp and biting, and smelling of plenty of volatile Salt: from whence we may safely conjecture that its use is to fix the blood, and . . . by volatilising the nervous juice, to take away the convulsive disposition.'[13] A hundred years later Dutch physician Pieter van Woensel (1747–1808) combated an outbreak of whooping cough among forty cadets with a similar 'decoction' of cup lichen mixed with a little mint syrup. All except one recovered.[14] *Cladonia* contains usnic acid, as does the hair lichen *Usnea*, used internally and for binding wounds. The name *Usnea* comes from the Arabic word *Ushnah*. Ebubekir Muhammed bin Zekeriya Razi (865–925) mentioned several medicinal uses in his *Liber Almansoris* (*Kitab al-Mansuri*), a text that was widely read in Europe.[15] The list of usnic acid's defensive capabilities is impressive: 'antibacterial, antifungal, antiviral, antioxidant, anti-inflammatory, antipyretic, anticancer, antigenotoxic, antimutagenic, antiplatelet, anti-ulcerogenic, and gastro-protective'.[16]

The use of skull lichens and skull moss – the names lichen and moss being used interchangeably at the time – to cure illness was definitely 'a curious chapter in the history of phytotherapy'.[17] They were in demand for the cure of epilepsy, nosebleeds and haemorrhaging. Lichens, probably *Parmelia saxatilis*, were scraped off human skulls collected from battlefields in Ireland and from the remains of the unburied. Skulls of those violently killed were preferred, because the vital spirits of the prematurely deceased were thought to have

Moss and lichen that grew on human skulls were used medicinally in 17th-century Europe. This woodblock print appeared in both Gerard's *Herball* of 1633 and Parkinson's *Theatrum botanicum* of 1640.

entered quickly into the growing material of the lichen, making it more powerful.[18] Skull lichen was also an important ingredient in a popular salve of the late Middle Ages known as *Unguentum armarium*, or weapon salve. Counterintuitively, the salve was applied to the weapon that caused the wound rather than the wound itself. By the mid-nineteenth century waning belief in the doctrine of signatures had undermined faith in lichens as medicinal agents. However, the efficacy of the lichens on Linnaeus' list and many others has been substantiated by modern methods of analysis and testing.

#lichendyes

Humans found in lichens a palette of dye colours with which to enhance their lives, perhaps a gift as valuable as their medicinal substances. In 1786 the Lyon Academy of Natural History's essay topic for its yearly prize was 'Which particular lichen species are of service in medicine and the arts?' Georg Franz Hoffmann (1760–1824), a doctor of medicine at the University of Erlangen, won with an essay about lichen dyes illustrated with 77 examples.[19] In 1805 Johan Peter Westring, physician to the king of Sweden as well as a botanist and lichenologist, published *The Color History of Swedish Lichens; or, The Way of Using Them for Dyeing and Other Household Purposes*.[20] Each of the 24 plates illustrates a different lichen with a colour chart displaying the shades available from that lichen. Not all lichens yield dyes, but many, even the most nondescript, can be made to produce innumerable shades of brown, green, orange, red, pink, purple and yellow. Dye lichens figure in the cultural history of indigenous peoples, cottage industries and turbulent empires.

The bright yellow of the wolf lichen (*Letharia vulpina*) appeared in crafts of Northern Europe's Sámi people and in the Pacific Northwest coast, where the presence of indigenous peoples dates back 10,000 years or more. Examples include the bold yellow stripes of the Chilkat dancing blankets made by the Tlingit living on the islands and coasts of southern Alaska, and the baskets dyed yellow by the Sinixt (also Sin Aixst), one of the First Nations of the Arrow Lakes Region of British Columbia. The Karuk living near the Klamath River, California, decorated their basket covers with porcupine quills stained yellow with wolf lichen dye.[21] While some lichens release pigments directly in boiling water, such as the wolf lichen's vulpinic acid, many lichen dyes are created by transforming lichen acids and other secondary compounds into dyes through soaking in ammonia, a process called ammonia fermentation. In the Scottish Highlands crofters made a lichen dye called crottle or crotal by scraping lichens, among them the skull lichen *Parmelia saxatilis*, off rocks with special 'crotal spoons' and fermenting them in urine, though the urine of beer drinkers was avoided as it was too weak. It is also said that Scottish fishermen avoided wearing garments dyed a misty brown because of the belief that 'what is taken from the rocks will return to the rocks'.[22]

Orchil lichens, a group of lichens that grew like weeds in the Mediterranean, became hugely important in the purple dye industry that began in antiquity. They yield reds and purples, depending on pH of the dye bath and other factors. Orchil is a byproduct of the colourless compound orcinol, a phenolic compound that breaks down into the dye orcein, or orchil, in the presence of oxygen and ammonia. Orcinol occurs in many lichens, but most famously in *Roccella tinctoria* and its relatives. They are fruticose lichens that grew plentifully on both sides of the Atlantic on sea cliffs. They went by numerous names in many languages but were often simply called the 'weeds'. Orchil adheres to silk and wool without a mordant, and it was also used to colour 'wood, marble, leather, wine, and foods'.[23]

Textile historian Dominique Cardon writes in her classic text *Natural Dyes: Sources, Traditions, Technology and Science* (2007) that the use of lichen purple began during the Akkadian empire of Mesopotamia (*c.* 2350–2150 BCE) and among the ancient Hebrews.[24] Purple became a colour much valued by Greek, Persian, Babylonian and Jewish rulers as a symbol of regality and wealth. Mention of a purple/red dye developed from scraping lichens off rocks appeared in the work of both Theophrastus (*c.* 300 BCE) and Pliny the Elder (*c.* 50 CE).[25] The story is complicated because there were two purples in antiquity: shellfish or mollusk purple and lichen purple. Archaeological evidence indicates that there were many dye centres producing shellfish purple in the ancient Mediterranean archipelago.[26] The Phoenicians (1500–300 BCE) produced Tyrian purple ('true purple') by extracting dye from the mucous glands of large predatory sea snails (*Murex*, *Thais*), a costly, labour-intensive and evil-smelling process that required 8,000 snails for 1 gram of dye.[27] Orchil (lichen purple) was thought to be a more brilliant, beautiful purple, but its colour was fugitive, while

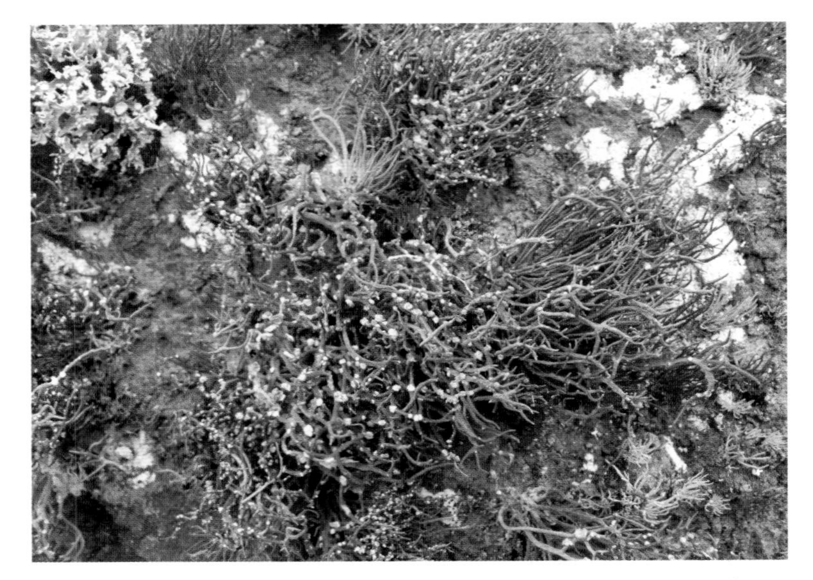

The orchil dye lichen *Roccella tinctoria* once grew like a weed on oceanside cliffs in many parts of the world, Canary Islands, Spain.

Tyrian purple was lightfast and more costly, a virtue considering that purple signified wealth. Researchers are still in the process of investigating historic textiles and artefacts because, as analytical chemist Chris Cooksey writes, 'Dyers have always been secretive to preserve their intellectual property.'[28]

Advanced methods of spectroscopic analysis have identified orchil in Ireland's *Book of Kells*, a Celtic gospel book created circa 800 CE; the *Vienna Genesis*, a sixth-century Greek manuscript found in Syria and stored at the Austrian National Library since 1664; and in silk textiles belonging to Vikings and Anglo-Saxons as well as artefacts found at burial sites in Norway and Denmark.[29] Shellfish purple disappeared at the end of the thirteenth century, while use of orchil became commercial in the fourteenth century when, according to a popular story, Italian merchant Frederigo (Ferro) carried orchil dye recipes from the Levant to Florence, where he established a lucrative, and secretive, monopoly employing many *tintori* (dyers). From the fifteenth to the nineteenth centuries the livelihoods of

'The Fall of Man', a folio from the *Vienna Genesis*, dyed with orchil lichen; the text was written in silver ink.

harvesters, traders, agents, brokers, chemists and dyers in many countries were dependent on the orchil dye industry. In Majorca orchil harvesters were called orxellers. It was a dangerous profession: harvesters dangled from ropes while tugging at lichens that were firmly attached by holdfasts to high cliffs. Further, spies pretending to be orxellers were hanged during the ongoing conflicts between Aragon and Majorca.[30] Antwerp served as the clearing house for vast amounts of unprocessed lichens, sometimes mixed with rocks

Collecting dye lichens was dangerous work in the Canary Islands, despite the insouciant appearance of Alfred Diston's *Gomero, Orchilla Gatherer*; 1828, watercolour.

to inflate yields, sent from the Spanish Canary Islands and the many Portuguese islands. It is estimated that 120 tons of orchil lichen were exported annually from the Canary Islands alone to England and Spain in the first half of the eighteenth century.[31] Cardon writes that:

> Joshua Wardle's dyeworks in Leek, Staffordshire, had more than one and a half tonnes of orchil in stock in December 1838! In addition to the pinks, reds, lilacs and violets known since the Middle Ages, it was now combined with a number of other dyestuffs to obtain various shades known by names such as stone, slate, sage green, bottle green, olives, fawns and a wide range of drabs, a colour favoured by English dyers.[32]

Demand for new sources led to coastal areas of North and South America (Baja California, Chile, Ecuador, Galápagos, Peru), Africa (Angola, Madagascar, Mozambique, Zanzibar), India and elsewhere.[33] Eventually orchil lichen populations crashed, and synthetic dyes arrived on the scene in the mid-nineteenth century. However, as of 2007 orchil lichens were still being collected for use in litmus, which was originally discovered in 1300 in Spain but primarily made from the sixteenth century onwards in the Netherlands.[34] Analysis of a family archive found in 2010 in a Devon attic, containing two centuries of records of a family of dye merchants in Leeds, revealed a picture of orchil dye commerce in England from the early 1800s onwards.[35] Textile artist Isabella Whitworth, one of the scholars who analysed the Bedford archive, writes, 'In 1877, a stock book records a total of 55 tons of various orchil lichens stored in the Leeds warehouses.'[36] She does not use lichen dyes in her work because, as she writes in her blog, 'lichens have suffered enough.'[37]

Now the conservation of dye lichens is self-monitored by hand-dyers, guided by Karen Diadick Casselman, who is both a historian and a practitioner. She writes in *Lichen Dyes: The New Source Book* (1996) that 'Lichen dyes are an education in ecology and ethics. To use lichens is to explore the point where craft and material culture intersect science and natural history. No other dyes provide a better opportunity to learn how to protect the environment.'[38] She also devised the Dyer's Code of Ethics, which includes buying a field guide to local and regional lichens, learning to identify at least five species by their

Pages from the research notebook of artist and orchil-dye historian Isabella Whitworth, who studies historical recipes for use by conservators, show shades from different dye baths.

Latin names and practising 'salvage botany' – only gathering those dislodged in some way, for example, by wind or by rock climbers, or found on fallen branches. Casselman also rescued Eileen Bolton's classic *Lichens for Vegetable Dyeing* (1960) from neglect.[39] A quest led her to a small weaver's cottage in Pont Eidda, North Wales, only to find that Bolton had died a year earlier. The daughter of artists, Bolton became a notable self-educated scholar, handweaver, hand-dyer and artist. Her colour plates render lichens in relation to more familiar plants for scale. Bolton objected to lichens being called 'humble little handmaidens of Nature'. Instead, she wrote: 'They are . . . grasping, tenacious and dogged . . . However carefully the first moon rocket is decontaminated, I am sure a tiny generative particle will find its way into the works . . . and on landing will begin to expand hopefully on a mountain of the moon.'[40]

#horticulturalmoss&lichen

Many gardeners only belatedly realize the horticultural potential of cryptogams in their gardens – usually they have moved in and made themselves permanent. George Schenk summarizes the story of the moss garden temples of Kyoto, Japan, this way: 'Mosses invited themselves into the gardens of Japan and thereby invented moss gardening.'[41] One of the most famous moss garden temples, Saihō-ji, now a World Heritage site, was founded in 731 by Emperor Shomu. In the fourteenth century Zen priest and garden designer Musō Soseki (Kokushi) (1275–1351) introduced gardening with stones, an aesthetic promoted in his 'Poem on Dry Mountain' whose opening lines read, 'without a speck of dust/ a high peak rises'.[42] When the moss temple fell into disrepair during the centuries that followed, moss moved in. Many of the ancient stones are now completely covered in moss. Schenk relishes the fact that mosses overwhelmed the 'stark agglomerations of stones', suggesting that 'there must have been a kind of communication between monks and mosses during the ritual reveries that are fundamental to the Zen faith.'[43] The rock-and-sand landscape of Ryōan-ji is another story. Also a UNESCO World Heritage site, it is an example of Kokushi's dry landscape or

Carpet mosses and whitewash-lichen-streaked tree trunks harmonize in Saihō-ji (Kokedera), a moss temple in Kyoto, Japan.

karesansui style, in which moss and lichen have not been allowed to proliferate to any great extent. Japanese garden historian François Berthier, who is a proponent of this style, writes:

> There is an overwhelming impression, initially, of sparse sterility – until one notices the moss that surrounds several of the rocks and the thin layer of lichen on some of them. Not much life for a garden, admittedly, but just enough, insofar as it provides a striking contrast to the unremittingly inorganic nature of the rest. In summer the bright green of the moss echoes the lush colors of the trees, while in winter its darker greens and mauves match the hues of both the evergreens and the branches of the deciduous trees that border the wall.[44]

It is not known exactly when mosses arrived at Saihō-ji and were allowed to stay by the mossmerized monks, but it has probably taken some hundreds of years for them to swallow the stones.

Home gardeners, however, can achieve pleasing results in their own lifetime. Co-creators of Sakonnet Garden in Little Compton, Rhode Island, John Gwynne and Mikel Folcarelli have developed their garden as a series of rooms, each with its own drama and aesthetic focus. Visitors enter through the moss room, where the dark trunks of huge old rhododendrons writhe like arboreal snakes over a peaceful carpet of moss. Gwynne and Folcarelli deliberately nurtured this mossy room to change visitors' sense of scale such that 'subsequently seen tree trunks and flowers [would] seem much larger than reality'.[45] Annie Martin, aka 'Mossin' Annie', experienced moss gardener and grower and author of *The Magical World of Moss Gardening* (2015), has a theory about 'moss magic' in the garden. She writes of its 'magnetic calm'. Though she grows and sells moss, much of her work is salvage botany. She collects moss doomed by logging and development and takes it back to her nursery. Conservation minded, she is a fervent moss partisan for reasons both aesthetic ('The petite leaves of *Climacium* species have a radiance that truly rivals fine emeralds') and biological ('The magnitude of millions of years of survival elevates them to a state of gentle giants').[46]

Lichens have recently attracted the attention of horticulturalists such as Margaret Roach, who wrote about 'lichen season' in the garden for the *New York Times*, and the editors of *Gardens Illustrated*, who featured a Finnish island garden where lichens took centre stage. Alison Pouliot writes in *Australian Garden History* that 'It is pretty hard to find a garden, especially an old one, without lichens . . . Lichens convert garden substrates into dappled tapestries, festoons of thalli, and carpeted dingles.'[47] While mosses in the garden may nourish mindfulness, lichens are, she notes, 'lavishly aesthetic', turning stones, benches and old rusty gates into works of art.

#moss&lichenart

In *Eye for Detail: Images of Plants and Animals in Art and Science, 1500–1630* (2017), cultural historian Florike Egmond closely examines the illustrations found in the *Libri Picturati*, a collection of unpublished natural history watercolours and drawings made in the sixteenth and seventeenth centuries and rediscovered in the twentieth century. Federico Cesi, founder of the Academy of the Lynx-Eyed (Accademia dei Lincei), Adriaen Coenen, a Dutch fish merchant, and Conrad Gessner, a Swiss polymath, made their realistic depictions for the wealthy Charles de Saint Omer of the Low Netherlands. Egmond writes that 'The Flemish *Libri Picturati* of the 1560s . . . provide evidence of a persistent fascination with "flowerless plants".'[48] Cesi made hundreds of drawings of cryptogams, whose flowerlessness implied they held secrets about reproduction and the origins of life. Egmond believes these observers, working premicroscope, were using their eyes to see microscopically, to *in*spect – to decipher the inner workings of these life forms – and that the microscope 'enabled naturalists to see what they had tried to imagine'.[49] Such imaginative *in*spection with the human eye continues today.

Watercolour *c.* 1560s showing several mosses (right) and lichens (left), from the *Libri Picturati*, a stunning collection of 16th-century natural history illustrations rediscovered in Poland after the Second World War.

In 2013 botanical illustrator Lizzie Harper posted 'Introduction to Lichens' on her website, an account of how she undertakes her portraits using knowledge gained as a member of the Institute of Analytical Plant Illustration.[50] In a 2016 post, 'Botanical Illustration: Step by Step Sphagnum Moss', she demonstrates how she starts with a pencil rough because 'the moss makes very little sense without shadows'. It was only when she removed her contact lenses – she is very near-sighted – that she could see the tiny details. Much of the challenge was to convey the topography of the sphagnum clump, which she did by applying several layers of washes. Representing the mercurial colours of sphagnum required additional washes.[51]

A 2016 show called 'Invisible Landscapes' featured the work of Sarah Hearn, who creates artificial lichen colony sculptures made up of photographs and drawings of specimens. She says that her sculptures are unplanned – they grow. In mixing lichens together she creates not just new art forms but new lichens. Prizes were given for finding twelve of her hidden-in-plain-sight lichen colonies, an activity intended to help people develop an eye for noticing lichens in their environment.[52] A 2021 exhibition titled 'We Are All Lichens' featured six artists whose hybrid compositions embodied 'recycling, transformation, and piecing together'. The phrase 'we are all lichens' is attributed to Scott Gilbert and was used by feminist scholar Donna Haraway in her provocative essay 'Tentacular Thinking' (2016).[53] Gilbert, who was a student of Haraway's in the history of science, envisions humans as holobionts: host organisms carrying multitudes of symbiotic organisms.[54]

In 'Lichen Artists and Artistic Lichenologists: Becoming What We Attend To' (2023) lichenologist artist Nastassja Noell profiles eleven lichen artists whose work has been informed by lichenological research. She writes, 'all of us who study lichens are perhaps lichenized in certain ways.'[55] One is reminded of Richard Spruce feeling himself becoming moss-like. She herself is now interested in doing transdisciplinary writing about the crossovers between lichen topics like dormancy (desiccation tolerance) and emergence (novelty) and human fields of study, such as de-growth economics and theories of the mind.[56]

#mosspeople

Nordic fairy tales and bryology come to life in the work of contemporary Finnish artist and ceramicist Kim Simonsson. In Germanic folklore there is a class of fairy folk named 'moss people', variously *moosleute* for 'moss folk', *holzleute* for 'wood folk' and *waldleute* for 'forest folk'. The *moosleute* of Germanic and Scandinavian folklore are moss-covered forest creatures with lichen hair, as described in this old ballad collected by Archibald Maclaren (1857):

> 'A moss-woman I' the hay-makers cry.
> And over the fields in terror they fly.
> She is loosely clad from neck to foot.
> In a mantle of Moss from the Maple's root

And like Lichen grey on its stem that grows
Is the hair that over her mantle flows.
Her skin, like the Maple-rind, is hard,
Brown and ridgy, and furrowed and scarred;
And each feature flat, like the bark we see,
Where a bough has been lopped from the bole of a tree,
When the newer bark has crept healingly round,
And laps o'er the edge of the open wound;
Her knotty, root-like feet are bare,
And her height is an ell from heel to hair.[57]

They were fickle, sometimes healing with medicinal herbs and sometimes bringing plague; needy – known for asking for sips of human breast milk; and temperamental – offerings of caraway bread caused angry fits. Finnish sculptor Kim Simonsson has created, almost by accident, a series of sculptures he calls 'Moss People'. Experimenting with a flocking technique he created a velvety surface on his ceramic sculptures. Its mossy green colour appeared accidentally when he overlaid yellow nylon flock to disguise a black pigment that he didn't like. He describes in a YouTube video how at about this time he discovered the beauty of mossy trees and forests upon moving from city to country. He realized 'Moss People' was a perfect name for his fairy-tale figures, who are often shown as solitary survivors in blasted landscapes, but the brightness of the moss children conveys the possibility of rejuvenation in the future.[58]

#moss&lichenwildlife

Many birds depend on mosses and lichens to ensure the survival of the next generation, and in the process have become 'skilled' bryologists and lichenologists, according to researchers. Nest-building is 'costly' for parents because gathering materials leaves them vulnerable to avian predators, but a high price is also paid by the chicks if the nests are not wholesome. Evidence that birds choose mosses selectively is relatively new, partly because mosses are hard to identify; ornithologists must become bryologists themselves or collaborate with them. In a 2018 study of three species of tits in Poland, researchers found that all three species favoured pleurocarps, the freely branching carpet-forming species, rather than the stiff, stocky acrocarps. Pleurocarps, growing as wefts, mats or pendants, lend themselves to being woven. Further, the largest tit chose the most 'robust' mosses for their heavier broods while the smaller tits chose the 'finest' moss species for their lighter broods.[59]

In 2020 Francisco E. Fontúrbel et al. presented a study that tells the story of the green-backed firecrown hummingbird (*Sephanoides sephanioides*), the sole hummingbird and the main pollinator of trees in the temperate rainforests of southern South America, which constructs 97 per cent of its nest from the least abundant moss in its environment.[60] Testing showed that the favoured moss, *Ancistrodes genuflexa*, contains five antipathogenic compounds that presumably increase the protective features of the nest, and hence chick survival. This moss is considered a 'cryptic' species: it makes up less than 1 per cent of moss biomass in the forest. It is now thought that mosses engineer important 'cryptic interactions' in temperate rainforests and boreal forests that are just now being discovered.[61]

Researchers agree that mosses provide insulation, improve sanitary conditions, maintain humidity, accommodate growing chicks through elasticity, and protect via antimicrobial, antifungal and antiparasitic compounds. More than fifty birds use mosses, including the red-necked grebe (sphagnum), the American bald eagle and the brown dipper, which collects aquatic mosses. The golden plover goes one step further, producing chicks whose birth down mimics their moss-lined nests. It nests in the Arctic tundra, where mosses and lichens thrive. The male prepares a 'scrape' in the ground, which the female lines with mosses and lichens. The chicks emerge covered in green-gold fluff, like the colour of their nest linings. They are 'precocial', able to walk and hunt for insects soon after birth, so they need camouflage immediately.[62]

While many hummingbirds construct nests with lichens, use by ruby-throated hummingbirds (*Archilochus colubris*) and the blue-grey gnatcatcher is said to be 'invariant'. They 'shingle' the outside of their nests with the rounded lobes of various shield lichens (*Flavoparmelia* spp., *Parmelia* spp. and *Punctelia* spp.), using spider silk as glue. They attach the nest to branches in the same way. Researchers wanted to know whether the hummingbirds were acting as dispersal agents. Observation is difficult because a ruby-throated hummingbird nest is approximately 5 centimetres wide and 2.5 centimetres tall; the gnatcatcher's nest is only slightly bigger. After three years a piece of powdery axil-bristle lichen (*Myelochroa aurulenta*) was still attached to the branch, indicating that hummingbirds may indeed be dispersing lichens from as far away as 40 metres. These cryptic interactions may have far-ranging effects: a global breeding population of more than 20 million ruby-throated hummingbirds is said to exist.[63]

The northern parula warbler (*Parula americana*) is an interesting case: it uses old man's beard lichen (*Usnea*) when it nests in Vermont but the lookalike flowering plant Spanish moss (*Tillandsia usneoides*)

A male blue-grey gnatcatcher (*Polioptila caerulea*) shingles its nest with lichens, providing camouflage, Millstone Aqueduct, West Windsor Township, New Jersey.

A collared Inca hummingbird (*Coeligena torquata fulgidigula*) improves nest hygiene with moss in the Ecuadorian Andes.

The nest of the common carder bee (*Bombus pascuorum*) with its waxy dome removed reveals collected moss.

when it nests in southern states. It is thought that this tiny warbler may retreat from breeding in Vermont as usnea there is threatened by air pollution from industrial parts of the Midwest.[64] The lace lichen (*Ramalina menziesii*) is also commonly used by birds.

Research ecologist Richard Broughton, who monitors the marsh tit population in Monks Wood, an ancient woodland in the UK, commented on X (13 March 2020), 'Fascinating how thrush nests outlive their original purpose for years, becoming arboreal moss gardens. Are the birds exploiting the moss for material, or is the moss exploiting the birds for dispersal, some kind of symbiosis?' Indeed, broad definitions of symbiosis do include such interactions.

Linnaeus named a beautiful ginger bumblebee, known for its velvety 'pilosity', the moss carder bee (*Bombus muscorum*; Latin *muscus*, moss) after its habit of gathering moss to cover its nest.[65] According to a nineteenth-century observer, moss carder bees line up single file in covered galleries with their backs towards the nest, passing along clumps of moss, which they card or comb of debris with their forelegs. Specialist builders at the nest site gather the moss and withered grass in a dome 7–10 centimetres above ground level and line it with wax.[66]

#moss&lichenconservation

The stories of rare mosses and lichens, and their disappearance and reappearance, enhance our understanding of the complexities of their biology. In some cases the race is to document species before they disappear; in others, it is to conserve. Bringing moss and lichen into botanic gardens in an intentional way is raising awareness about cryptogams. The Royal Botanic Garden Edinburgh started the first cryptogamic garden in 1992. The idea took hold in India, where the harvesting of lichens for medicines and dyes has caused conservation concern. Renewed interest in herbal drugs led to 'exhaustive' exploitation of lichens in the Himalayas such that few propagules are left for renewal of populations. Researchers note that 'To save a rainforest species you probably need a rainforest,' but creating lichen gardens is one way to save and conserve some of the diversity.[67]

Sometimes research itself is a danger to a rare moss. The entire known population of the Antarctic endemic moss *Schistidium deceptionense* has been endangered by trampling as a by-product of research. *Schistidium*, with twelve species, is described as 'the richest moss genus in the Antarctica'. Bryophytes in the Antarctic often live in areas of geothermal activity, which are known to have sustained life forms during glaciation cycles. *Schistidium deceptionense* was first described in 2003 from the summit ridge of Caliente Hill on the volcanic island of Deception in the South Shetland Islands archipelago, in one of these areas. The issue of bryophyte damage on Caliente Hill was presented to the Antarctic Treaty Consultative Meeting (ATCM) in 2014 and 2015, with a proposed code of conduct for these highly sensitive areas.[68] Bryologists are keeping a close eye on the ancient

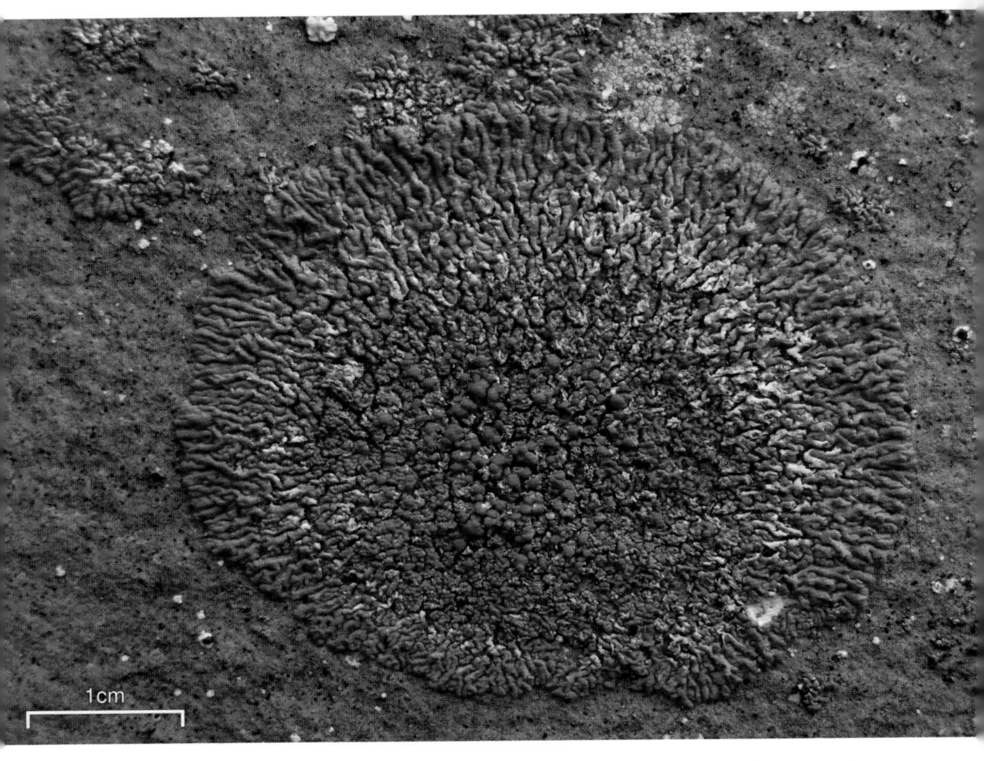

The elegant sunburst lichen (*Rusavskia elegans*, formerly *Xanthoria elegans*) returned from a lengthy stay in outer space ready to resume growth.

moss beds of east Antarctica where four decades of increasing dryness and cold were followed by warming (beginning in 2016), which resulted with a 'mind-boggling' heat wave in 2022.[69]

Bryophyte inventory projects are under way worldwide to maintain local records or extend current records; often new species are described in the process. One such inventory is taking place on the rugged, high-elevation Beartooth Plateau in Wyoming, where the 'globally rare' *Philonotis yezoana* was found far from its usual habitats – the oceanic temperate area of East Asia and North America.[70] Sometimes a rare species, thought to be extinct locally, reappears. The stunning golden-eye lichen (*Teloschistes chrysophthalmus*) recently appeared at the National Trust's Golden Cap property near Lyme

Regis in Dorset. Reported extinct in Britain, it is thought to have recolonized from spores blown from Brittany. Its reappearance after a hundred-year absence is a mystery.[71]

Quite a different lichen, the rock gnome lichen (*Cetradonia linearis*) of the southern Appalachians is one of two on the endangered species list in the USA. It is a narrow endemic in an area that is threatened by climate change, resource extraction and invasive tree pests. The balsam woolly adelgid threatens high-elevation spruce fir forests, and the hemlock woolly adelgid threatens hemlock forests at lower elevations. A lover of constant fog, its status is further threatened by reports of reduced cloud immersion of the high elevations. It is an odd lichen, occurring in patches of little green fingers (squamules).[72]

The case of the 'exceptionally rare' Cornish path-moss (*Ditrichum cornubicum*) demonstrates other challenges in conservation. First discovered in 1963, it is known only from three former copper mines in Cornwall, UK, and County Cork, Ireland. One of a group of 'copper mosses', it is a metallophyte that thrives on bare soil rich in arsenic, copper and cobalt, but declines quickly from competition with larger, similarly tolerant bryophytes and lichens. At Phoenix United Mine near Bodmin Moor, Cornwall, turf-stripping to reduce competition resulted in a twenty-fold increase in the population. The physiological means by which the moss tolerates these phytotoxins still needs further research. In February 2019 poet Ella Frears and artist Ben Sanderson presented an evening dedicated to the Cornish path-moss. They were joined by students from Falmouth School of Art, who presented moss-inspired work in an event sponsored by Back from the Brink (2017–21), a programme dedicated to saving the UK's most threatened species.[73]

In 2000 Ricardo Rozzi invited attendees at a celebration of the opening of the Beagle Channel between Chile and Argentina to look at a moss-covered rock with magnifying glasses, initiating the 'ecotourism with a hand-lens movement'.[74] Soon after, the UNESCO Cape Horn Biosphere Reserve, an area with over 1,000 species of mosses and lichens, was established. One of the authors of *Miniature*

Forests of Cape Horn: Ecotourism with a Hand Lens (2012), Rozzi sees 'miniature forest' and 'hand lens' as metaphors that 'amplify' our vision of small organisms.[75] He writes about how a 'change of lenses' – from 'observe' to 'conserve' to 'cultivation' of bioethics – helps us see, and think, better. Mosses and lichens are becoming more visible, recognized for their extraordinary biology and beauty. One remembers the line from Lew Welch's poem, 'why am I made to kneel and peer at Tiny?' Because that's how we learn to respect mosses and lichens for their many contributions to the well-being of the Earth and its inhabitants.

Timeline

c. 2.7 billion years ago	Cyanobacteria (a lichen photobiont) begin oxygenating the planet (the Great Oxygenation Event)
c. 1.5 billion years ago	Fungi (a lichen mycobiont) diverge from other life forms
c. 1 billion years ago	Green algae (another lichen photobiont) evolve
c. 350–287 BCE	Theophrastus describes a lichen resembling *Usnea* in *Historia plantarum*: 'The substance which some call tree-moss and which resembles rags is borne only by the aigilops (Turkey-oak); it is grey and rough and hangs down for a cubit's length, like a long shred of linen' (§ 3.8.5, www.topostext.org)
1175	Al-Razi's *Liber Almansoris*, promoting many uses for Ushnah (the lichen *Usnea*), is translated into Latin
1400–1850	Orchil lichens (*Roccella tinctoria* et al.) are harvested on both sides of the Atlantic for the purple dye industry
1596	Li Shizhen mentions use of TuMaZang or Horse-Mane-of-the Earth (*Polytrichum*) as a diuretic in his *Compendium of Materia Medica*, a use still considered effective
1663	Robert Hooke presents his microscopical observations of the 'common moss' to the Royal Society of London
1741	*Historia muscorum* by Johannes Jacob Dillenius appears in print

1753	Linnaeus groups mosses and lichens as cryptogams in *Species plantarum*
1774	Johann Hedwig visualizes moss sperm ('Kügelchen')
1798	*Lichenographiae svecicae prodromus* by Erik Acharius appears in print
1801	*Species muscorum frondosorum* by Johann Hedwig appears in print
1843	William Starling Sullivant travels through the Allegheny Mountains with Asa Gray, producing the *Musci Alleghaniensis* in two volumes (1845, 1846)
1851	Wilhelm Hofmeister links the life cycle of moss with that of higher plants, introducing the concept of an alternation of generations
1867	Simon Schwendener proposes the dual nature of lichens in Rheinfelden, Switzerland
1868	Frances Elizabeth Tripp popularizes mosses with the publication of *British Mosses: Their Homes, Aspects, Structure, and Uses*
1877	Albert Bernhard Frank uses the term 'symbiosis' to describe the fungal-algal relationship in lichens
1896	Coslett Herbert Waddell establishes the Moss Exchange Club, renamed the British Bryological Society in 1923
1898	Elizabeth Gertrude Knight Britton and Abel Joel Grout establish the Sullivant Moss Society, which becomes the American Bryological and Lichenological Society in 1970
1907	Nina L. Marshall popularizes mosses and lichens in the USA with the publication of *Mosses and Lichens*
1914	Edinburgh surgeon Charles Walker Cathcart initiates the collection of sphagnum moss to be used for wound dressings in the First World War

1917	Moss- and lichen-like fossils found in the Rhynie chert beds of Aberdeen, Scotland
1921	*Lichens* by Annie Lorrain Smith appears in print, quickly becoming a classic work in lichen literature
1958	Thomas Douglas (Dougal) Victor Swinscow (1917–1992) founds the British Lichen Society with 23 other participants
1965	*The Bog People* by P. V. Glob popularizes the study of bodies found in sphagnum bogs
1967	The newly formed International Association for Lichenology offers a formal definition of the lichen
1994	Saihō-ji (Kokedera, or 'moss temple') in Kyoto, Japan, is registered as a UNESCO World Heritage Site
1998	Rosmarie Honegger publishes article titled 'The Lichen Symbiosis – What Is So Spectacular About It?', urging biologists to bring the study of lichens into mainstream biology (*The Lichenologist*, XXX/3, pp. 193–212)
2005	Lichen-like fossils estimated to be 600 million years old are found in marine sediments of the Doushantuo Formation, South China
2008	Trevor Goward publishes 'Nameless Little Things', the first of twelve invited essays on lichens (*Evansia*, XXV/3, pp. 23–5)
2014	Scientists revive a moss frozen for 1,500 years in Antarctica
2014	The lichen *Xanthoria elegans* maintains viability after one and a half years in outer space
2015	Governor Jerry Brown designates the lace lichen (*Ramalina menziesii*) as the California State Lichen
2016	Toby Spribille et al. announce the discovery of basidiomycete yeasts in the cortex of ascomycete lichens (*Science*, CCCLIII/6298, pp. 488–92)

2018	Study of alluvial mudrock suggests moss-like plants engineered mud 480 million years ago
2018	Study of peat-moss-like (*Sphagnum* sp.) fossils found in Ordovician deposits (Wisconsin, USA) indicate peat-moss lineage is 460–607 million years old
2020	David L. Hawksworth and Martin Grube redefine lichens as complex ecosystems (*New Phytologist*, CCXXVII/5, pp. 1281–3)
2020	Bord na Móna (The Peat Board) phases out peat harvesting in Ireland and reassigns workers to bog reclamation
2022	UK bans the of sale of peat moss to amateur gardeners from 2024 onwards
2022	Toby Spribille et al. review the 'goods-and-services exchange' between fungus and photobiont and suggest new approaches to lichen research (*New Phytologist*, CCXXXIV/5, pp. 1566–82)
2023	David Eldridge et al. publish a global field study quantifying contributions of soil mosses: they cover 9.4×106 km^2 (an area the size of China or the USA); add nitrogen, phosphorus and magnesium to the soil and reduce soil-borne plant pathogens beneath them; and more (*Nature Geoscience*, XVI/5, pp. 430–38)
2023	Swedish researchers document the decline of the flagship species *Usnea longissima*, the world's longest lichen, in both a national park and managed forests (*Forest Ecology and Management*, DXLVI (15 October 2023), article 121369)

References

Introduction: The Cryptogamic Carpet

1 Rachel Carson, *The Sea around Us* [1950] (New York, 2018), pp. 6–7.
2 Lincoln Taiz and Lee Taiz, *Flora Unveiled: The Discovery and Denial of Sex in Plants* (New York, 2017), p. 465.
3 Irwin M. Brodo, Sylvia Duran Sharnoff and Stephen Sharnoff, *Lichens of North America* (New Haven, CT, and London, 2001).
4 Kathy Merrifield and Lynn Royce, 'Invertebrate Communities in Mosses', www.yaquina.info, accessed 27 November 2023.
5 Wolfgang Elbert et al., 'Contribution of Cryptogamic Covers to the Global Cycles of Carbon and Nitrogen', *Nature Geoscience*, 5 (2012), pp. 459–62, available at www.nature.com; Ulrich Pöschl, 'Wallflowers of the Earth System', www.mpg.de, 1 June 2012.
6 Christopher Cardona-Correa et al., 'Peat Moss-Like Vegetative Remains from Ordovician Carbonates', *International Journal of Plant Sciences*, CLXXVII/6 (2016), available at www.journals.uchicago.edu.
7 Michael S. Ignatov and Elena Maslova, 'Fossil Mosses: What Do They Tell Us about Moss Evolution?', *Bryophyte Diversity and Evolution*, XLIII/1 (2021), pp. 72–97; William J. McMahon and Neil S. Davies, 'Evolution of Alluvial Mudrock Forced by Early Land Plants', *Science*, CCCLIX/6379 (2018), pp. 1022–4.
8 Rosmarie Honegger, Dianne Edwards and Lindsey Axe, 'The Earliest Records of Internally Stratified Cyanobacterial and Algal Lichens from the Lower Devonian of the Welsh Borderland', *New Phytologist*, CXCVII/1 (2012), pp. 264–75.
9 Xunlai Yuan et al., 'Lichen-Like Symbiosis 600 Million Years Ago', *Science*, CCCVIII/5724 (2005), pp. 1017–20.
10 Matthew P. Nelsen et al., 'The Macroevolutionary Dynamics of Symbiotic and Phenotypic Diversification in Lichens', *Proceedings of the National Academy of Sciences*, CXVIII/35 (2020), pp. 21495–503.
11 Jen-Pan Huang et al., 'Accelerated Diversifications in Three Diverse Families of Morphologically Complex Lichen-Forming Fungi Link to

Major Historical Events', *Scientific Reports*, IX/8918 (2019), available at www.nature.com; 'When the Dinosaurs Died, Lichens Thrived', www.fieldmuseum.org, 27 June 2019.

12 '*Rhizocarpon geographicum* (L.) DC.', http://species.nbnatlas.org, accessed 27 November 2023.

1 Curious Vegetation

1 'John Bartram to Mark Catesby', *c.* March 1741, in *The Correspondence of John Bartram, 1734–1777*, ed. Edmund Berkeley and Dorothy Smith Berkeley (Gainesville, FL, 1992), p. 152; see also William Darlington, *Memorials of John Bartram and Humphry Marshall* (Philadelphia, PA, 1849).

2 Quoted in Marc J. Ratcliff, *The Quest for the Invisible: Microscopy in the Enlightenment* (Farnham, 2009), p. 51.

3 Hilda Grieve, *A Transatlantic Gardening Friendship, 1694–1777* (Saffron Walden, 1981), pp. 5–27.

4 Stephanie Volmer, '"Taste," "Curiosity," and the Letters of John Bartram and Peter Collinson', in *America's Curious Botanist: A Tercentennial Reappraisal of John Bartram 1699–1777*, ed. Nancy E. Hoffman and John C. Van Horne (Philadelphia, PA, 2004), p. 76.

5 Dawson Turner, ed., *Extracts from the Literary and Scientific Correspondence of Richard Richardson* (Yarmouth, 1835), p. vii.

6 'Letter XXVII, Feb. 12, 1702–3(?), Mr. Vernon to Mr. Richardson', ibid., p. 73.

7 B. M. Buddle, 'Moss-Cropper Extraordinaire: The Rev. Adam Buddle (1662–1715)', *The Linnaean*, XXIV/4 (2008), p. 14.

8 Ibid., pp. 13–19.

9 'Vernon, William (*c.* 1666–1711)', https://plants.jstor.org, accessed 27 November 2023.

10 'Letter XXVII', in *Extracts*, ed. Turner, pp. 74–5.

11 John Jac. Dillenius, *Historia muscorum* (London, 1678); see also 'Historia Muscorum', https://herbaria.plants.ox.ac.uk, accessed 27 November 2023. Oxford University maintains a website where the text of *Historia muscorum* is linked to his herbarium.

12 'Letter LX, Feb. 28, 1718–19, Dr. Sherard to Dr. Richardson', in *Extracts*, ed. Turner, p. 146.

13 Dillenius, *Historia muscorum*, p. 10.

14 Quoted in Grieve, *Transatlantic*, pp. 10–11.

15 See www.bartramsgarden.org, accessed 27 November 2023.

16 Grieve, *Transatlantic*, p. 18.

17 'Peter Collinson: 1694–1768', www.quakersintheworld.org, accessed 27 November 2023.

18 Volmer, '"Taste, "Curiosity,"', p. 76.

19 William R. Buck and Elizabeth P. McLean, '"Mosses" in Lord Petre's Herbarium Collected by John Bartram', *Bartonia*, 51 (1985), pp. 17–33.

20 Mark Lawley, 'The Influence of Social Background on the Emergence of British Field Botanists in the 17th–19th Centuries: William Wilson, a Case Study', *Field Bryology*, 94 (February 2008), pp. 28–35.
21 'Hobson's *Musci Britannici*', one of 101 Treasures of Chetham's, https://library.chethams.com, accessed 27 November 2023.
22 James Cash, *Where There's a Will, There's a Way! Or, Science in the Cottage; An Account of the Labours of Naturalists in Humble Life* [1873] (Cambridge, 2011), p. 94.
23 Anne Secord, 'Artisan Botany', in *Cultures of Natural History*, ed. N. Jardine, J. A. Secord and E. C. Spary (Cambridge, 1996), pp. 378–93.
24 Quoted ibid., p. 382.
25 Ibid.
26 John Percy, 'Scientists in Humble Life: The Artisan Naturalists of South Lancashire', *Manchester Region History Review*, V/I (1991), pp. 3–10.
27 Secord, 'Artisan Botany', p. 381.
28 Ratcliff, *Quest*, p. 51.
29 Lincoln Taiz and Lee Taiz, *Flora Unveiled: The Discovery and Denial of Sex in Plants* (New York, 2017), p. 323.
30 Ratcliff, *Quest*, p. 51.
31 Ibid.
32 Ibid., p. 49.
33 Klaus Schillinger, 'Johann Gottfried Kohler – Inspector at the Mathematical-Physical Salon in Dresden – an Active Observer of the Starry Sky in the Last Quarter of the 18th Century', *Analytica Chimica Acta*, 33 (2007), pp. 257–300; Ratcliff, *Quest*, p. 308 (*fl.* stands for floruit, flourished, used when birth and death dates are unknown).
34 Ratcliff, *Quest*, p. 167.
35 Harald Sack, 'Johann Hedwig – the Father of Bryology', http://scihi.org, 8 December 2017.
36 John Farley, *Gametes and Spores: Ideas about Sexual Reproduction, 1750–1914* (Baltimore, MD, 1982), p. 91.
37 Ratcliff, *Quest*, pp. 237–8.
38 Ibid., p. 467.
39 'Hedwig, Johann', www.encyclopedia.com, accessed 27 November 2023.
40 David Brewster, ed., *The Edinburgh Encyclopaedia*, vol. XII (Edinburgh, 1830), p. 730, available at https://babel.hathitrust.org.
41 Ibid.
42 Ibid., p. 731.
43 Michelle J. Price, 'Johannes Hedwig and the Founding of Modern Bryology', Invited Contribution, Golden Oldies Symposium, IAB IMOSS SEB 2019 Conference, www.bryology2019.com, accessed 27 November 2023.
44 Sack, 'Johann Hedwig'.
45 Quoted in Frank Teichmann, 'The Emergence of the Idea of Evolution in the Time of Goethe', trans. Jon McAlice, www.waldorflibrary.org, accessed 27 November 2023.

46 A. G. Morton, *History of Botanical Science* (London, 1981), pp. 293–4.
47 Ibid., p. 295.
48 Ratcliff, *Quest*, p. 51.
49 Quoted in Morton, *History*, p. 353.
50 Arthur Henry Church, *Thalassiophyta and the Subaerial Transmigration*, Oxford Botanical Memoirs 3 (Oxford, 1919), p. 4.
51 Morton, *History*, p. 399.
52 Donald R. Kaplan and Todd J. Cooke, 'The Genius of Wilhelm Hofmeister: The Origin of Causal-Analytical Research in Plant Development', *American Journal of Botany*, LXXXIII/12 (1996), pp. 1647–60.
53 F. O. Bower, *The Origin of a Land Flora: A Theory Based upon the Facts of Alternation* (London, 1908), p. 1.

2 Moss: Versatile Minimalist

1 Robert Hooke, *Micrographia* [1665] (New York, 1961), p. 131.
2 P. W. Richards, 'Robert Hooke on Mosses', *Occasional Papers of the Farlow Herbarium of Cryptogamic Botany*, 16 (June 1981), pp. 137–46, available at www.biodiversitylibrary.org.
3 R. Medina et al., 'Evolutionary Dynamism in Bryophytes: Phylogenomic Inferences Confirm Rapid Radiation in the Family Funariaceae', *Molecular Phylogenetics and Evolution*, CXX (2018), pp. 240–47.
4 M.C.F. Proctor, 'Mosses and Alternative Adaptation to Life on Land', *New Phytologist*, CXLVIII/1 (2000), pp. 1–3.
5 Ron Porley and Nick Hodgetts, *Mosses and Liverworts* (London, 2005), p. 67.
6 Janice M. Glime, 'Adaptive Strategies: Travelling the Distance to Success', in *Bryophyte Ecology*, vol. I, ebook (Houghton, MI, 2017), Ch. 4–8.
7 Lily R. Lewis et al., 'First Evidence of Bryophyte Diaspores in the Plumage of Transequatorial Migrant Birds', *PeerJ*, 2 (2014), e424.
8 Des A. Callaghan et al., 'Long-Term Survival of Bryophytes Underground: An Investigation of the Diaspore Bank of *Physcomitrium eurystomum* Sendtn.', *Journal of Bryology*, XLIV/3 (2022), pp. 208–16; see images on X @CallaghanDes: 15 March 2020 and 12 December 2022.
9 Anton Kerner von Marilaun, *The Natural History of Plants*, vol. I, trans. and ed. F. W. Oliver [1890–91] (London, 1904), p. 385.
10 Yoan Coudert et al., 'Multiple Innovations Underpinned Branching Form Diversification in Mosses', *New Phytologist*, CCXV/2 (2017), pp. 840–50; Barbara J. Crandall-Stotler and Sharon E. Bartholomew-Began, 'Morphology of Mosses (Phylum Bryophyta)', www.flora.huh.harvard.edu, accessed 22 January 2024.
11 Catherine La Farge-England, 'Growth Form, Branching Pattern, and Perichaetial Position in Mosses', *The Bryologist*, XCIX/2 (1996), pp. 170–86.
12 Coudert et al., 'Multiple Innovations', p. 842.
13 Ibid.
14 Ibid.

15 Heino Lepp, 'Leaves', *Australian Bryophytes*, www.anbg.gov.au, accessed 27 November 2023.
16 See Bill Malcolm and Nancy Malcolm, *Mosses and Other Bryophytes: An Illustrated Glossary* (Nelson, New Zealand, 2000), an excellent resource for moss vocabulary with photographs and illustrations.
17 Stephanie Stuber, 'The World of Mosses', *Arnoldia*, LXXI/1 (2013), pp. 26–35.
18 James Shevock, 'The Amazing Design of a Moss Leaf', *BryoString*, 3 (2015), pp. 9–19, available at https://bryophyte.cnps.org.
19 'Can You Recognize These Three Common Churchyard Mosses?', www.greenchristian.org.uk, accessed 27 November 2023.
20 'Grey-Cushioned Grimmia – *Grimmia pulvinata*', www.naturespot.org.uk; '*Grimmia pulvinata*', www.efloras.org; Brian Ecott, 'Common Wall Mosses', www.hainaultforest.co.uk, all accessed 27 November 2023.
21 See Malcolm and Malcolm, *Mosses*, for vocabulary and definitions.
22 Shevock, 'Amazing Design'.
23 '*Rosulabryum subtomentosum*', www.nzplants.auckland.ac.nz, accessed 7 April 2020. The University of Auckland in New Zealand maintains a digital photographic database showing the life histories of this species and many others.
24 Karl B. McKnight et al., *Common Mosses of the Northeast and Appalachians* (Princeton, NJ, and Oxford, 2013), p. 211.
25 Jerry Jenkins, *Mosses of the Northern Forest: A Photographic Guide* (Ithaca, NY, 2020), p. 164.
26 Michael Lüth, *Mosses of Europe: A Photographic Flora*, www.milueth.de, accessed 27 November 2023.
27 Jennifer Frazer, 'How Mosses Have Sex in Spite of Their Swimming-Challenged Sperm', *Scientific American*, www.blogs.scientificamerican.com, 27 July 2012.
28 Todd N. Rosenstiel et al., 'Sex-Specific Volatile Compounds Influence Microarthropod-Mediated Fertilization of Moss', *Nature*, 489 (2012), pp. 431–3.
29 Janice M. Glime and Irene Bisang, 'Sexuality: Size and Sex Difference', in *Bryophyte Ecology*, vol. I, ebook (Houghton, MI, 2017), Ch. 3–3; Frida Rosengren and Nils Cronberg, 'The Adaptive Background of Nannandry: Dwarf Male Distribution and Fertilization in the Moss *Homalothecium lutescens*', *Biological Journal of the Linnean Society*, CXIII/1 (2014), pp. 74–84.
30 M. Van der Velde et al., 'The Reproductive Biology of *Polytrichum formosum*: Clonal Structure and Paternity Revealed by Microsatellites', *Molecular Ecology*, X/10 (2021), pp. 2423–34.
31 Jessica M. Budke, 'Exploring Calyptra Function: A Dissertation Saga in Summary', http://mossplants.fieldofscience.com, 25 July 2013.
32 A. J. Grout, *Mosses with Hand-Lens and Microscope* [1903] (Ashton, MD, 1965), p. 32.

33 Ibid., p. 29.
34 Derek Christie, '*Tortula muralis* – My First Moss', www.microscopy-uk.org.uk, accessed 27 November 2023.
35 See Matt Candeias,'The Peculiarly Tiny World of *Buxbaumia* Mosses', www.indefenseofplants.com, 10 February 2019; 'Draft Management Recommendations for Green Bug Moss *Buxbaumia viridis* (DC.) Moug. & Nestl.', Bureau of Land Management (1996), www.bim.gov, accessed 27 November 2023.
36 '*Buxbaumia aphylla* Hedw.', www.dnr.state.mn.us, accessed 27 November 2023.
37 Ameline Guillet, Vincent Hugonnot and Florine Pépin, 'The Habitat of the Neglected Independent Protonemal Stage of *Buxbaumia viridis*', *Plants*, X/1 (2021), p. 83.
38 Robin Wall Kimmerer, 'Portrait of Splachnum', in *Gathering Moss: A Natural and Cultural History of Mosses* (Corvallis, OR, 2003), p. 121.
39 Monica Suleiman and Andi Maryani A. Mustapeng, 'The Discovery of Dung-Loving Moss *Tayloria octoblepharum* (Splachnaceae) on "Toilet Pitchers" in Borneo', *Malayan Nature Journal*, LXXI/1 (2019), pp. 17–20; 'A Poop-Loving Moss Discovered Living on Poop-Eating Pitcher Plants', www.indefenseofplants.com, 1 July 2019.
40 Paul Marino et al., 'The Ecology and Evolution of Fly Dispersed Dung Mosses (Family Splachnaceae): Manipulating Insect Behaviour through Odour and Visual Clues', *Symbiosis*, XLVII/2 (2009), pp. 61–76.
41 Joan Bingley, 'Dung Mosses', www.quekett.org, accessed 27 November 2023.
42 Bjørnar Kjensli, 'Lemmings' Loss Is Bounty for Moss', www.sciencenorway.no, 23 March 2012.
43 '*Splachnum rubrum* Hedw.', www.dnr.state.mn.us, accessed 27 November 2023.
44 M. L. Gonzalez et al., '*In vitro* Micropropagation and Long-Term Conservation of the Endangered Moss *Splachnum ampullaceum*', *Biologia Plantarum*, L/3 (2006), pp. 339–45.
45 K. Mägdefrau, 'Life-Forms of Bryophytes', in *Bryophyte Ecology*, ed. A.J.E. Smith (Dordrecht, 1982), pp. 45–58.
46 Brent Mishler, 'The Biology of Bryophytes, with Special Reference to Water', *Fremontia*, XXXI/3 (2002), pp. 34–8.
47 Jenkins, *Mosses*, p. 16.
48 See Melinda Waterman's image of moss turf in 'The Amazing Antarctic Moss', www.science.org.au, accessed 27 November 2023.
49 See Jerry Jenkins's 'Moss Lesson 1: Walking the Tracks', www.northernforestatlas.org, 11 May 2020.
50 Alison Haynes, 'Silver Moss Is a Rugged Survivor in the City Landscape', *The Conversation*, www.theconversation.com, 22 March 2019; R. Jia et al., 'Antagonistic Effects of Drought and Sand Burial Enable the Survival of the Biocrust Moss *Bryum argenteum* in an Arid Sandy Desert', *Biogeosciences*, XV/4 (2018), pp. 1161–72.

51 'Silvery Bryum Moss – *Bryum argenteum*', https://fieldguide.mt.gov, accessed 27 November 2023; '*Bryum argenteum*', http://floranorthamerica.org, accessed 27 November 2023.
52 Zane Raudenbush et al., 'Divergence in Life-History and Developmental Traits in Silvery-Thread Moss (*Bryum argenteum* Hedw.) Genotypes Between Golf Course Putting Greens and Native Habitats', *Weed Science*, LXVI/5 (2018), pp. 642–50.
53 Richards, 'Robert Hooke on Mosses'.
54 Kathrin Rousk, Davey L. Jones and Thomas H. DeLuca, 'Moss-Cyanobacteria Associations as Biogenic Sources of Nitrogen in Boreal Forest Ecosystems', *Frontiers in Microbiology*, IV (2013), available at www.frontiersin.org.
55 Herman Melville, *Moby Dick* [1851] (New York, 1926), p. 134.

3 Lichen: Complex Individuality

1 Ed Yong, 'The Overlooked Organisms That Keep Challenging Our Assumptions about Life', *The Atlantic*, www.theatlantic.com, 17 January 2019.
2 Albert Schneider, *A Guide to the Study of Lichens* (Boston, MA, 1898), p. 1.
3 Annie Lorrain Smith, 'Introduction', in *Lichens* (Cambridge, 1921), available at www.gutenberg.org.
4 Yong, 'Overlooked Organisms'.
5 Ibid.
6 Charles C. Plitt, 'A Short History of Lichenology', *The Bryologist*, XXII/6 (1919), p. 78.
7 W. Lauder Lindsay, *A Popular History of British Lichens* (London, 1856), p. 22.
8 Plitt, 'Short History', p. 79.
9 Irwin M. Brodo, Sylvia Duran Sharnoff and Stephen Sharnoff, *Lichens of North America* (New Haven, CT, and London, 2001), p. 8.
10 A. Thell et al., eds, 'In the Footsteps of Erik Acharius', 20th biennial meeting of the Nordic Lichen Society, Vadstena, 11–15 August 2013. For online version, see www.up.lub.lu.se, accessed 1 January 2022.
11 Smith, *Lichens*.
12 M. E. Mitchell, '"Such a Strange Theory": Anglophone Attitudes to the Discovery That Lichens Are Composite Organisms, 1871–1890', *Huntia*, XI/2 (2002), pp. 193–207.
13 'Species: *Collema tenax*', www.fs.usda.gov, accessed 22 January 2024.
14 Mitchell, '"Such a Strange Theory"', p. 195.
15 Rosmarie Honegger, 'Great Discoveries in Bryology and Lichenology – Simon Schwendener (1829–1919) and the Dual Hypothesis of Lichens', *The Bryologist*, CIII/2 (2000), p. 308.
16 Plitt, 'Short History', p. 81.
17 Mitchell, '"Such a Strange Theory"', p. 196.
18 Ibid., p. 197.
19 Plitt, 'Short History', p. 82.

20 Ibid.
21 Frederic E. Clements, 'The Polyphyletic Disposition of Lichens', *American Naturalist*, XXXI/364 (1897), pp. 277–84.
22 J. M. Crombie, 'On the Algo-Lichen Hypothesis', *Botanical Journal of the Linnean Society*, XXI/135 (1884), p. 282.
23 Smith, *Lichens*.
24 Bradford D. Martin and Ernest Schwab, 'Symbiosis: "Living Together" in Chaos', *Studies in the History of Biology*, IV/4 (2012), pp. 7–25.
25 Ibid., p. 8.
26 Ibid.
27 Ibid.
28 See 'What's in a Lichen? How Scientists Got It Wrong for 150 Years', *National Geographic*, www.youtube.com, 25 January 2018.
29 Stuart Crawford, 'Ethnolichenology of *Bryoria fremontii*: Wisdom of Elders, Population Ecology, and Nutritional Ecology', MSc Thesis, Interdisciplinary Studies, University of Victoria, British Columbia, Canada, 2007.
30 Ibid.
31 Toby Spribille et al., 'Basidiomycete Yeasts in the Cortex of Ascomycete Macrolichens', *Science*, CCCLIII/6298 (2016), pp. 488–92; Ed Yong, 'How a Guy from a Montana Trailer Park Overturned 150 Years of Biology', *The Atlantic*, www.theatlantic.com, 21 July 2016; Marina Richie, 'Overturning 150 Years of Science', *Medium*, www.medium.com, 9 March 2017.
32 Toby Spribille, personal communication, 3 December 2023.
33 Elizabeth Pennisi, 'A Lichen Ménage à Trois', *Science*, CCCLIII/6298 (2016), p. 337.
34 David L. Hawksworth and Martin Grube, 'Lichens Redefined as Complex Ecosystems', *New Phytologist*, CCXXVII/5 (2020), pp. 1281–3.
35 Gregor Pichler et al., 'How to Build a Lichen: From Metabolite Release to Symbiotic Interplay' (Tansley Review), *New Phytologist*, CCXXXVIII/4 (2023), p. 1363.
36 Lynn Margulis and Eva Barreno, 'Looking at Lichens', *BioScience*, LIII/8 (2003), pp. 776–8.
37 Trevor Goward, 'I. Face in the Mirror', in *The Book: Twelve Readings on the Lichen Thallus* (2012), pp. 1–3, www.waysofenlichenment.net, accessed 1 August 2023.
38 Pichler et al., 'How to Build a Lichen'.
39 Martin Grube and Toby Spribille, 'Exploring Symbiont Management in Lichens', *Molecular Ecology*, XXI/13 (2012), pp. 3098–9.
40 William B. Sanders, 'Complete Life Cycle of the Lichen Fungus *Calopadia puiggarii* (Pilocarpaceae, Ascomycetes) Documented In Situ: Propagule Dispersal, Establishment of Symbiosis, Thallus Development, and Formation of Sexual and Asexual Reproductive Structures', *American Journal of Botany*, CI/11 (2014), pp. 1836–48.
41 Brodo, Duran Sharnoff and Sharnoff, *Lichens*, p. 8.

42 Trevor Goward, 'XII: Formal Propositions', in *The Book: Twelve Readings on the Lichen Thallus* (2012), p. 34, www.waysofenlichenment.net, accessed 1 August 2023.
43 Hadi Nazem-Bokaee et al., 'Towards a Systems Biology Approach to Understanding the Lichen Symbiosis: Opportunities and Challenges of Implementing Network Modelling', *Frontiers in Microbiology*, XII (2021), available at www.frontiersin.org.
44 Maria Grimm et al., 'The Lichens' Microbiota, Still a Mystery?', *Frontiers in Microbiology*, XII (2021), available at www.frontiersin.org.
45 Toby Spribille et al., '3D Biofilms: In Search of the Polysaccharides Holding Together Lichen Symbioses', *FEMS Microbiology Letters*, CCCLXVII/5 (2020), available at www.ncbi.nim.nih.gov.
46 Grimm et al., 'The Lichens' Microbiota'.
47 Ursula Goodenough, Ralf Wagner and Robyn Roth, 'Lichen 4. The Algal Layer', *Algal Research*, LVIII (2021), available at www.sciencedirect.com.
48 Rosmarie Honegger, 'Questions about Pattern Formation in the Algal Layer of Lichens with Stratified (Heteromerous) Thalli', *Bibliotheca Lichenologica*, 25 (1987), pp. 59–71.
49 Trevor Goward, 'III. Credo', in *The Book: Twelve Readings on the Lichen Thallus* (2012), p. 4, www.waysofenlichenment.net, accessed 1 August 2023.
50 Toby Spribille et al., 'Evolutionary Biology of Lichen Symbioses', *New Phytologist*, CCXXXIV/5 (2022), pp. 1566–82.
51 Sanders, 'Complete Life Cycle', p. 1847.
52 Pichler et al., 'How to Build a Lichen', p. 1364.
53 Grube and Spribille, 'Exploring Symbiont 'Management', p. 3099.
54 Sergio Pérez-Ortega et al., 'Lichen Myco- and Photobiont Diversity and Their Relationships at the Edge of Life (McMurdo Dry Valleys, Antarctica)', *FEMS Microbiology Ecology*, LXXXII/2 (2012), pp. 429–48.
55 Leonardo M. Casano et al., 'Two *Trebouxia* Algae with Different Physiological Performances Are Ever-present in Lichen Thalli of *Ramalina farinacea*. Coexistence versus Competition?', *Environmental Microbiology*, XIII/3 (2011), pp. 806–18.
56 Goward, 'XII. Formal Propositions', p. 49.
57 Merlin Sheldrake, *Entangled Life: How Fungi Make Our Worlds, Change Our Minds and Shape Our Futures* (New York, 2020), p. 75.
58 Richard Armstrong and Tom Bradwell, 'Growth of Crustose Lichens: A Review', *Geografiska Annaler*, XCII/1 (2010), pp. 3–17.
59 See images posted on X by April Windle (@aprilwindle) of a lobarion in Argyll, Scotland, 13 January 2022.
60 Oliver Gilbert, *Lichens* (London, 2000), p. 81.
61 'Charity Attempts Largest Transplant of Ancient and Rare Lichen in Efforts to Protect Its Future', www.insideecology.com, 23 November 2020.
62 M.-M. Kytöviita and P. D. Crittenden, 'Growth and Nitrogen Relations in the Mat-Forming Lichens *Stereocaulon paschale* and *Cladonia stellaris*', *Annals of Botany*, C/7 (2007), pp. 1537–45.

63 Abigail Robison et al., 'Fruticose Lichen Communities at the Edge: Distribution and Diversity in a Desert Sky Island on the Colorado Plateau', *Conservation*, II/4 (2022), available at www.mdpi.com.
64 Goward, 'III. Credo', p. 2.
65 Sheldrake, *Entangled Life*, p. 88.
66 Ursula Goodenough, 'Introduction to the Lichen Ultrastructure Series', *Algal Research*, 51 (2020), available at www.sciencedirect.com.
67 Ibid.
68 Honegger, 'Great Discoveries', p. 311.

4 Cosmopolitan Extremophiles

1 M. R. Turetsky et al., 'The Resilience and Functional Role of Moss in Boreal and Arctic Ecosystems' (Tansley Review), *New Phytologist*, CXCVI/I (2012), pp. 49–67.
2 Brent Mishler, 'The Biology of Bryophytes, with Special Reference to Water', *Fremontia*, XXXI/3 (2002), p. 38.
3 Ilse Kranner et al., 'Desiccation-Tolerance in Lichens: A Review', *The Bryologist*, CXI/4 (2008), p. 587.
4 Melvin J. Oliver, Zoltán Tuba and Brent D. Mishler, 'The Evolution of Vegetative Desiccation Tolerance in Land Plants', *Plant Ecology*, CLI/I (2000), pp. 85–100.
5 Melvin J. Oliver, Jeff Velten and Brent D. Mishler, 'Desiccation Tolerance in Bryophytes: A Reflection of the Primitive Strategy for Plant Survival in Dehydrating Habitats?', *Integrative and Comparative Biology*, XLV/5 (2005), pp. 788–99.
6 Kranner et al., 'Desiccation-Tolerance', p. 589.
7 Francisco Gasulla et al., 'Advances in Understanding of Desiccation Tolerance of Lichens and Lichen-Forming Algae', *Plants (Basel)*, X/4 (2021), available at www.ncbi.nlm.nih.gov.
8 Ibid.
9 Ibid.
10 Oliver Gilbert, *Lichens* (London, 2000), pp. 169–87.
11 'Australian Lichens: Ecology', www.anbg.gov.au, accessed 27 November 2023.
12 William Purvis, *Lichens* (London, 2000), p. 31.
13 See Gilbert, *Lichens*, p. 80, for lichen bark preferences and Jerry Jenkins, *Mosses of the Northern Forest: A Photographic Guide* (Ithaca, NY, and London, 2020) for descriptions of moss habitat preferences.
14 Jerry Jenkins, 'Moss Lesson 1 – Walking the Tracks', www.northernforestatlas.org, 11 May 2020.
15 Ethel Mellor, 'Lichens and Their Action on the Glass and Leadings of Church Windows', *Nature*, CXII (1923), pp. 299–300; 'Vitricolous Lichens', www.anbg.gov.au, accessed 27 November 2023.
16 Kris Sales, Laurie Kerr and Jessie Gardner, 'Factors Influencing Epiphytic Moss and Lichen Distribution within Killarney National Park', *Bioscience*

Horizons: The International Journal of Student Research, IX (2016), available at www.academic.oup.com.

17 J. L. Gressit, J. Sedlacek and J.J.H. Szent-Ivany, 'Flora and Fauna on Backs of Large Papuan Moss-Forest Weevils', *Science*, CL/3705 (1965), pp. 1833–5.

18 Walter Fertig, 'Ten Things You Might Not Know About Lichens, But Wish You Did', https://biokic.asu.edu, accessed 27 November 2023.

19 Janice M. Glime, 'Water Relations: Leaf Strategies – Structural', in *Bryophyte Ecology*, vol. I, ebook (Houghton, MI, 2017), Ch. 7–4a.

20 Kranner et al., 'Desiccation-Tolerance'.

21 'Freshwater and Coastal Lichens', www.nhm.ac.uk, accessed 27 November 2023.

22 Kerry Knudsen, Jana Kocourková and Bruce McKune, '*Sarcogyne mitziae* (Acarosporaceae), a New Species from Biotic Soil Crusts in Western North America', *The Bryologist*, CXVI/2 (2013), pp. 122–6.

23 Chen Na, ed.,'Desert Moss Found to Be Expert of Water Collection for Survival', https://english.cas.cn, 12 July 2016.

24 See photographs in 'Desiccation and Diversity in Dryland Mosses: *Syntrichia caninervis*', https://3dmoss.berkeley.edu, accessed 27 November 2023.

25 Volkmar Wirth, *Lichens of the Namib Desert* (Göttingen, 2010).

26 Frank Bungartz and Volkmar Wirth, '*Buellia peregrina* sp. nov., a New, Euendolithic Calcicolous Lichen Species from the Namib Desert', *The Lichenologist*, XXXIX/1 (2007), pp. 41–5.

27 Lloyd R. Stark et al., 'Do the Sexes of the Desert Moss *Syntrichia caninervis* Differ in Desiccation Tolerance? A Leaf Regeneration Assay', *International Journal of Plant Sciences*, CLXVI/1 (2005), pp. 22–9.

28 Sabrina Imler, 'This Moss Uses Quartz as a Parasol', *New York Times*, www.nytimes.com, 29 July 2020; Jenna T. B. Ekwealor and Kirsten M. Fisher, 'Life Under Quartz: Hypolithic Mosses in the Mohave Desert', *PLoS One*, https://journals.plos.org, 22 July 2020.

29 Jenny Davis et al., 'Evolutionary Refugia and Ecological Refuges: Key Concepts for Conserving Australian Arid Zone Freshwater Biodiversity Under Climate Change', *Global Change Biology*, XIX/7 (2013), pp. 1970–84.

30 Zhao Pan et al., 'The Upside-Down Water Collection System of *Syntrichia caninervis*', *Nature Plants*, 2 (2016), available at www.nature.com.

31 'This Moss Has Developed the Ultimate Water Collection Toolkit', *Science Daily*, www.sciencedaily.com, 6 June 2016.

32 Xiaozhe Chen et al., 'Three-Dimensional Maskless Fabrication of Bionic Unidirectional Liquid Spreading Surfaces Using a Phase Spatially Shaped Femtosecond Laser', ACS *Applied Material Interfaces*, XIII/11 (2021), pp. 13781–91; Michael Franco, 'Physics Whizzes Create "Black Hole" to Stop Urinal Splash-back', www.cnet.com, 30 November 2015.

33 Ellen McHale, '7 Interesting Things about Mosses', www.kew.org, 23 May 2020.

34 Beatriz Fernández-Marín et al., 'Symbiosis at Its Limits: Ecophysiological Consequences of Lichenization in the Genus *Prasiola* in Antarctica', *Annals of Botany*, CXXIV/7 (2019), pp. 1211–26.
35 'Moss Study Evaluates Climate Change Impact on Arctic Ecosystems', www.cordis.europa.eu, accessed 27 November 2023.
36 Marc Oliva and Jesús Ruiz-Fernández, *Past Antarctica: Paleoclimatology and Climate Change* (London, 2020).
37 Damien Ertz and Andre Aptroot, 'Lichens from the Utsteinen Nunatak (Sør Rondane Mountains, Antarctica), with the Description of One New Species and the Establishment of Permanent Plots', *Phytotaxa*, CXCI/1 (2014), pp. 99–114.
38 Edwin B. Bartram, 'The Second Byrd Antarctic Expedition: Botany. III. Mosses', *Annals of the Missouri Botanical Garden*, XXV/2 (1938), pp. 719–24; Serena Zaccara et al., 'Multiple Colonization and Dispersal Events Hide the Early Origin and Induce a Lack of Genetic Structure of the Moss *Bryum argenteum* in Antarctica', *Ecology and Evolution*, X/16 (2020), pp. 8959–75.
39 Fernandez-Marin et al., 'Symbiosis at Its Limits'.
40 Maria Grimm et al., 'The Lichens' Microbiota, Still a Mystery?', *Frontiers in Microbiology*, XII (2021), available at www.frontiersin.org.
41 Sergio Pérez-Ortega et al., 'Hidden Diversity of Marine Borderline Lichens a New Order of Fungi: *Collemopsidiales* (*Dothideomyceta*)', *Fungal Diversity*, LXXX (2016), pp. 285–300.
42 Scott Trimble, 'A Tale of Two Lichens: Adaptations to Extreme Climate', www.cid-inc.com, 18 April 2022.
43 Janice M. Glime, 'Adaptive Strategies: Growth and Life Forms', in *Bryophyte Ecology*, vol. I, ebook (Houghton, MI, 2017), Ch. 4–5.
44 Nell Greenfieldboyce, 'Herd of Fuzzy Green "Glacier Mice" Baffles Scientists', www.npr.org, 22 May 2020; Theresa Machemer, 'Herds of Moss Balls Mysteriously Roam the Arctic Together', *Smithsonian Magazine*, www.smithsonianmag.com, 2 June 2020.
45 Jón Eyþórsson, *Journal of Glaciology*, I/9 (1951), p. 503.
46 J. H. Kidder, 'The Natural History of Kerguelen Island', *American Naturalist*, X/8 (1876), pp. 481–4.
47 Anita Sanchez, 'Glacier Mouse', www.anitasanchez.com, 18 November 2017.
48 S. J. Coulson and N. G. Midgley, 'The Role of Glacier Mice in the Invertebrate Colonization of Glacial Surfaces: The Moss Balls of the Falljökull, Iceland', *Polar Biology*, XXXV (2012), pp. 1651–8.
49 See photographer Andy Murray's website 'A Chaos of Delight' for images, www.chaosofdelight.org, accessed 17 July 2021.
50 Kathy Merrifield and Lynn Royce, 'Invertebrate Communities in Mosses', www.yaquina.info, accessed 27 November 2023.
51 Dianne Edwards, 'New Insights into Early Land Ecosystems: A Glimpse of a Lilliputian World', *Review of Palaeobotany and Palynology*, XC/3–4 (1996), pp. 159–74.

52 'A Chaos of Delight', www.achaosofdelight.org, accessed 17 July 2021.
53 David Copestake, 'Moss Mites in Woodlands', www.woodlands.co.uk, 8 February 2013.
54 Heather T. Root et al., 'Arboreal Mite Communities on Epiphytic Lichens of the Adirondack Mountains of New York', *Northeastern Naturalist*, XIV/3 (2007), pp. 425–38; Stef Bokhorst, 'Lichen Physiological Traits and Growth Forms Affect Communities of Associated Invertebrates', *Ecology*, XCVI/9 (2015), pp. 2394–407.
55 Janice M. Glime, 'Arthropods: Mites (Acari)', in *Bryophyte Ecology*, vol. II, ebook (Houghton, MI, 2017), Ch. 9-1.
56 Uri Gerson, 'Bryophytes and Invertebrates', in *Bryophyte Ecology*, ed. A.J.E. Smith (Dordrecht, 1982), pp. 291–331.
57 D. R. Nelson, 'The Hundred-Year Hibernation of the Water Bear', *Natural History*, LXXXIV/7 (1975), pp. 62–5.
58 William Randolph Miller, 'Tardigrades', *American Scientist*, www.americanscientist.org, accessed 27 November 2023.
59 Gerson, 'Bryophytes', p. 298.
60 Esme Roads, Royce E. Longton and Peter Convey, 'Millennial Timescale Regeneration in a Moss From Antarctica', *Current Biology*, XXIV/6 (2014), available at www.cell.com.
61 Aaron L. Gronstal, 'Lichens Can Survive Space Conditions for Extended Periods', www.phys.org, 22 October 2014.

5 Bogland

1 Howard Crum, *A Focus on Peatlands and Peat Mosses* (Ann Arbor, MI, 1988), p. 111.
2 Karen Russell, 'The Bog Girl', *New Yorker*, 13 June 2016.
3 Quoted in Håkan Rydin and John K. Jeglum, with Aljosja Hooijer, *The Biology of Peatlands* (Oxford, 2006), p. 274.
4 Paul Simons, 'Plantwatch: Is Sphagnum the Most Underrated Plant on Earth?', *The Guardian*, www.theguardian.com, 15 January 2019.
5 David Segal, 'Who Will Profit from Saving Scotland's Bogs?', *New York Times*, www.nytimes.com, 5 May 2022.
6 Rydin and Jeglum, with Hooijer, *Biology of Peatlands*, p. 65.
7 See Edward Struzik, *Swamplands: Tundra Beavers, Quaking Bogs, and the Improbable World of Peat* (Washington, DC, 2021); Annie Proulx, *Fen, Bog and Swamp: A Short History of Peatland Destruction and Its Role in the Climate Crisis* (New York, 2022).
8 A. Jonathan Shaw et al., 'Range Change Evolution of Peat Mosses (*Sphagnum*) within and between Climate Zones', *Global Change Biology*, XXV/1 (2019), pp. 108–20.
9 'Sphagnum', www.cpbr.gov.au; 'Getting Started with Sphagnum', www.ohiomosslichen.org; 'Sphagnum Bogs', www.cronodon.com; 'Sphagnopsida: Morphology', www.ucmp.berkeley.edu, all accessed 27 November 2023.

10 Håkan Rydin, Urban Gunnarsson and Sebastian Sundberg, 'The Role of *Sphagnum* in Peatland Development and Persistence', in *Boreal Peatland Ecosystems,* vol. CLXXXVIII: *Ecological Studies*, ed. R. K. Wieder and D. H. Vitt (Berlin, 2006), pp. 47–65.
11 See www.etymonline.com, accessed 27 November 2023.
12 James Prosek, 'A Botanist in Swedish Lapland', *New York Times*, www.nytimes.com, 16 May 2017.
13 Struzik, *Swamplands*, p. 20.
14 'What Do We Know about Peruvian Peatlands?', www.cifor.org, accessed 27 November 2023.
15 Struzik, *Swamplands*, p. 20; Crum, *Focus*, p. 153.
16 Crum, *Focus*, p. 11.
17 David Smith, 'Peat Bog as Big as England Found in Congo', *The Guardian*, www.theguardian.com, 27 May 2014; Frederick C. Draper et al., 'The Distribution and Amount of Carbon in the Largest Peatland Complex in Amazonia', *Environmental Research Letters*, IX/12 (2014), Article 124017; 'Congo Peat: The "Lungs of Humanity" Which Are Under Threat', BBC *News*, www.bbc.com, 16 June 2022.
18 '12 of the Most Mysterious Forests in the World', https://blog.tentree.com, accessed 27 November 2023.
19 Robert Maxwell, 'Essay on Moss', in *The Natural and Agricultural History of Peat-Moss or Turf-Bog, Etc*, ed. Andrew Stelle (Edinburgh, 1826), p. 184, digital version available at www.google.com/books.
20 Dianne Meredith, 'Hazards in the Bog: Real and Imagined', *Geographical Review*, XCII/3 (2002), pp. 319–32.
21 Michelle Z. Donahue, 'The Mad Dash to Figure Out the Fate of Peatlands', *Smithsonian Magazine*, www.smithsonianmag.com, 20 April 2016.
22 J. W. Hotson, 'Sphagnum Used as a Surgical Dressing in Germany during the World War', *The Bryologist*, XXIV/6 (1921), pp. 89–96.
23 Meredith, 'Hazards'; Robert Lloyd Praeger, *The Way That I Went* [1937] (Cork, 2014), pp. 373–5.
24 'Moving Bog Wipes Out a Village in Roscommon, Ireland', *The Call*, XCVII/31 (31 December 1904), California Digitial Newspaper Collection, available at www.cdnc.ucr.edu.
25 'Flows to the Future End of Conference Project', www.theflowcountry.org.uk, 12 September 2019.
26 'Sphagnum', www.cpbr.gov.au.
27 'Moss', www.theflowcountry.org.uk, 12 September 2019; Ruth Maclean, 'What Do the Protectors of Congo's Peatlands Get in Return?', *New York Times*, www.nytimes.com, 21 February 2022.
28 'List of Bogs', www.wikipedia.com, accessed 27 November 2023.
29 Peter Moore, 'Totally Mired: The Most Beautiful Bogs from around the World', www.wanderlust.com, 25 October 2019.
30 Fred Pearce, 'Climate Warning as Siberia Melts', *New Scientist*, www.newscientist.com, 10 August 2005.

31 Emily Toner, 'The Secret World of Life (and Death) in Ireland's Peat Bogs', *New York Times*, www.nytimes.com, 19 October 2019.
32 For images of leaf morphology, see 'Sphagnum Bogs', www.cronodon.com.
33 Crum, *Focus*, p. 153.
34 On 12 January 2023 an X user posted a video of a rotifer swimming in the shelter of a moss leaf.
35 Rydin and Jeglum, with Hooijer, *Biology of Peatlands*, p. 84.
36 Victor L. Mironov, Aleksei Y. Kondratev and Anna V. Mironova, 'Growth of Sphagnum Is Strongly Rhythmic: Contribution of the Seasonal, Circalunar and Third Components', *Physiologia Plantarum*, CLXVIII/4 (2020), pp. 765–76.
37 Neil Bell, *The Hidden World of Mosses* (Edinburgh, 2023), p. 147.
38 Robert Braithwaite, *The Sphagnaceae or Peat-Mosses of Europe and North America* (London, 1880), p. 6, digital version available at www.google.com/books.
39 Mark Lawley, 'Bygone Bryologists: Robert Braithwaite (1824–1917)', *Field Bryology*, 91 (2007), pp. 26–8; Heinjo J. During, Betty Verduyn and Bart F. van Tooren, 'On the Increase of *Trematodon ambiguus* in Lowland Belgium and the Netherlands', *Lindbergia*, XXXI/3 (2007), pp. 101–8; Sam Bosanquet, 'Vagrant Epiphytic Mosses in England and Wales', *Field Bryology*, 107 (2012), pp. 3–17.
40 Wolfgang Hofbauer, James H. Dickson and Ronald D. Porley, '*Trematodon ambiguus* (Hedw.) Hornsch. (Musci: Bruchiaceae) Reported from Northern Tyrol, Austria: Spore Germination and Protonemal Development', Naturwissenschaftlich Medizinischer Vereins (Innsbruck), July 2011, pp. 35–44, available at www.zobodat.at.
41 Quoted in Howard Crum, *Structural Diversity of Bryophytes* (Ann Arbor, MI, 2001), p. 176.
42 Dwight L. Whittaker and Joan Edwards, '*Sphagnum* Moss Disperses Spores with Vortex Rings', *Science*, CCCXXIX/5990 (2010), p. 406.
43 Janice M. Glime, 'Adaptive Strategies: Spore Dispersal Vectors', in *Bryophyte Ecology*, vol. I, ebook (Houghton, MI, 2017), Ch. 4–9.
44 Crum, *Focus*, p. 133.
45 Ibid., p. 113.
46 Christian Fritz et al., 'Sphagnum Mosses – Masters of Efficient N-Uptake While Avoiding Intoxication', *PLoS One*, IX/1 (2014), e79991.
47 Crum, *Focus*, p. 62.
48 Ibid., p. 133.
49 Robert Zottoli, 'Field Trip through a Typical Maine Raised Bog (Peatland)', www.rzottoli.wordpress.com, accessed 27 November 2023.
50 Braithwaite, *Sphagnaceae*, p. 11.
51 Hans Joosten and Donal Clarke, *Wise Use of Mires and Peatlands: Background and Principles Including a Framework for Decision-Making* (Devon, 2002).
52 Segal, 'Who Will Profit'; 'Welcome to Wildland', www.wildland.scot, accessed 27 November 2023.

6 Literary Ecology

1 Herman Melville, *Moby Dick* [1851] (New York, 1926), p. 556.
2 Mick Herron, *Slow Horses* (New York, 2010), p. 13.
3 Robin Wall Kimmerer, *Gathering Moss: A Cultural and Natural History of Moss* (Corvallis, OR, 2003).
4 Elizabeth Gilbert, *The Signature of All Things* (New York, 2013).
5 Elizabeth Gilbert, 'Me and My Moss Mentor!', www.elizabethgilbert.com, 7 October 2013.
6 Claudia Dreifus, 'Elizabeth Gilbert Finds Inspiration behind the Garden Gate', *New York Times*, www.nytimes.com, 4 November 2013.
7 Gilbert, *Signature*, p. 446.
8 'Dicranum', www.wikipedia.org, accessed 27 November 2023.
9 Sean R. Edwards, *Mosses in English Literature*, British Bryological Society (Cardiff, 1992).
10 Ibid., p. 2.
11 Ibid.
12 Ibid., p. 4.
13 Ann Radcliffe, '*A Journey Made in the Summer of 1794* . . .', (1795), available at www.fadedpage.com, accessed 27 November 2023.
14 Dorothy Wordsworth, *Recollections of a Tour Made in Scotland* [1874] (New Haven, CT, 1997), p. 64.
15 Kate Flint, 'Ruskin and Lichen', www.courtauld.ac.uk, accessed 27 November 2023.
16 See for example 'A Walk in the Forest, and a Long Love Affair with Lichen', www.livingwittily.typepad.com, 29 June 2020.
17 P. G. Wodehouse, *Right Ho, Jeeves* (London, 1934), p. 7.
18 William Shakespeare, *The Comedy of Errors* (Act II, Scene II, l. 177), see Alfred Harbage, ed., *The Complete Pelican Shakespeare* (Baltimore, MD, 1969).
19 J.R.R. Tolkien, *The Return of the King* [1955] (New York, 1991), pp. 973–4.
20 G. K. Chesterton, 'On Mr. Rudyard Kipling and Making the World Small', from 'Heretics', www.pagebypagebooks.com, accessed 27 November 2023.
21 'G. K. Chesterton', www.wikipedia.com, accessed 27 November 2023.
22 George Bernard Shaw, *Misalliance: A Debate in One Sitting* (New York, 1957), p. 85, digital version available at www.google.com/books.
23 'Alexander Pope and the Eighteenth-Century Garden', www.c18media.wordpress.com, 11 November 2012.
24 Alexander Pope, 'Epistle I to Sir Richard Temple, Lord Cobham' (1734), ll. 15–24, p. 6, *Eighteenth Century Collections Online*, www.quod.lib.umich.edu, accessed 27 November 2023.
25 Alexander Pope, 'The Dunciad Variorum', Book IV, ll. 449–50 (1743), *Eighteenth Century Collections Online*, www.quod.lib.umich.edu, accessed 27 November 2023.
26 Madeline Miller, *Circe* (New York, 2018), p. 21.
27 Beatrix Potter, *The Tale of Timmy Tiptoes* (London, 1911), p. 32.
28 Tove Jansson, *The Summer Book* [1972] (New York, 2008), pp. 12–14.

29 Lucy Knight, '"A Masterpiece": Why Tove Jansson's *The Summer Book* Is as Relevant as Ever at 50', *The Guardian*, www.theguardian.com, 1 September 2022; Susannah Clapp, 'How Tove Jansson's Love of Nature Shaped the World of the Moomin', *The Guardian*, www.theguardian.com, 5 June 2021; Ali Smith, 'Lights on a String', *The Guardian*, www.the guardian.com, 12 July 2003.
30 It was transcribed from Taimie Boswell, a Romani, on 9 January 1915, in Oswaldtwistle, Lancashire, by Katharine M. Briggs and Ruth L. Tongue, and appeared in their book *Folktales of England* (Chicago, IL, 1965).
31 Angela Carter, ed., *The Old Wives' Fairy Tale Book* (New York, 1990), p. 49.
32 Philip Pullman, 'The Fairytale of Mossycoat', *The Guardian*, www.theguardian.com, 11 October 2009.
33 Carter, ed., *Old Wives*, p. ix.
34 Donald A. Mackenzie, ed., *Scottish Wonder Tales from Myth and Legend* [1917] (Mineola, NY, 1997), p. 11.
35 Ibid., 65.
36 Kimmerer, *Gathering*, p. 10.
37 John Cowper Powys, *A Glastonbury Romance* (Woodstock, NY, 1996), pp. 512–13.
38 Quoted in Paul Richards, *A Book of Mosses* (Edinburgh, 1950), p. 20.
39 Pascal Quignard, *Mysterious Solidarities* (*Les solidarités mystérieuses*) [2011], trans. Chris Turner (Calcutta, 2021), p. 82.
40 Ibid., p. 208.
41 Ibid., p. 223.
42 Ibid., p. 187.
43 Forrest Gander, *Twice Alive: An Ecology of Intimacies* (New York, 2019), p. 5.
44 Hillary Rosner, 'In a Place for the Dead, Studying a Seemingly Immortal Species', *New York Times*, www.nytimes.com, 31 December 2012.
45 Gander, *Twice Alive*, p. 9.
46 Ibid.
47 Ibid., p. 17.
48 Lew Welch, 'Springtime in the Rockies, Lichen', broadside [Cranium Press] (San Francisco, CA, 1971).
49 Philip Whalen, 'Commonplace Discoveries: Lew Welch', www.thenewblackbartpoetrysociety.wordpress.com, 31 August 2016.
50 Richard J. Nevle and Steven Nightingale, *The Paradise Notebooks: 90 Miles across the Sierra Nevada* (Ithaca, NY, 2022), pp. 134–7.
51 Brenda Hillman, *Extra Hidden Life, among the Days* (Middletown, CT, 2019).
52 Ibid., p. 56.
53 Ibid., p. 57.
54 Ibid., p. 58.
55 Marilynne Robinson, *Jack* (New York, 2020), p. 26.
56 Tim Robinson, *Stones of Aran: Pilgrimage* [1986] (New York, 2008), p. 168.
57 Piter Kehoma Boll, 'Friday Fellow: Black Wart Lichen', www.earthlingnature.wordpress.com, 29 October 2021.
58 Lydia Davis, *Our Strangers* (London, 2023), pp. 192–9.

59 Helen Macdonald, *H Is for Hawk* (New York, 2014), p. 11.
60 Nikita Arora, 'The Many Meanings of Moss', *The Guardian*, www.theguardian.com, 3 November 2022.
61 Ibid.
62 Bill Bryson, *A Short History of Nearly Everything* (New York, 2003), p. 336.
63 Robert Macfarlane and Stanley Donwood, *Ness* (London, 2019).
64 Henry David Thoreau, '5 February 1853', *The Journal of Henry David Thoreau* (New York, 1962), p. 488.

7 Curious Observers: A Field Trip

1 F. A. Pottle, ed., *Boswell's London Journal, 1762–1763* (New York, 1950), p. 38.
2 William Lauder Lindsay, *A Popular History of British Lichens* (London, 1856), pp. 8–9.
3 Mako Nazu and Brian Thompson, 'What's behind Japan's Moss Obsession?', *The Conversation*, www.theconversation.com, 10 December 2015.
4 For example, see the obituary of the engaging Alan Crundwell: A. G. Perry and D. G. Long, 'Alan Cyril Crundwell, B.Sc. (1923–2000)', *Journal of Bryology*, XXIII/3 (2001), pp. 267–72.
5 Asa Gray, 'Memoir of William Starling Sullivant 1803–1873', in *National Academy of Science Biographical Memoirs*, vol. I, pp. 277–85, www.nasonline.org, accessed 27 November 2023; Annie Morrill Smith, 'William Starling Sullivant. January 15, 1803–April 30, 1873', *The Bryologist*, VIII/1 (1905), pp. 1–3.
6 Betsy Butler, 'My William Starling Sullivant Treasure Hunt Had All the Elements of a Thrilling Page Turner', www.beesfirstappearance.wordpress.com, 11 June 2013.
7 Quoted in 'Sketch of William Starling Sullivant', *Popular Science Monthly*, 48 (March 1896), http://en.wikisource.org, accessed 27 November 2023.
8 Mark R. D. Seaward, 'William Borrer (1781–1862), Father of British Lichenology', *The Bryologist*, CV/1 (2002), pp. 70–77.
9 Ibid., p. 73.
10 Anonymous, 'In Memory of the Late William Borrer, Esq., F.R.S., F.L.S., of Henfield', *New Phytologist*, 6 (1861–63), pp. 70–83.
11 Ibid., p. 71.
12 William Edward Nicholson, 'William Mitten: A Sketch with Bibliography', *The Bryologist*, X/1 (1907), pp. 1–5.
13 Des Callaghan, 'Typification and Diagnosis of *Weissia* × *mittenii* (Bruch & Schimp.) Mitt. emend. A.J.E.Sm. (*Weissia multicapsularis* × *W. rostellata*) (Pottiaceae, Bryophyta)', *Journal of Bryology*, XLI/3 (2019), pp. 243–8.
14 Brad Scott, 'William Mitten, Hurstpierpoint and the Bryophytes of the World', *Field Bryology*, 122 (2019), p. 28.
15 Ibid., pp. 27–34.
16 Vivienne Manchester, 'Memories of Hurst', 1975, privately published, www.sussexpostcards.info, accessed 27 November 2023.

17 Scott, 'William Mitten', p. 34.
18 Ron Porley and Nick Hodgetts, *Mosses and Liverworts* (London, 2005), p. 341.
19 E. M. Holmes, 'Obituary Notice and Bibliography of William Mitten, ALS (1819–1906)', *Proceedings of the Linnean Society of London*, 119–21 (1906–7), pp. 49–54.
20 Scott, 'William Mitten', pp. 27–34; Brad Scott, 'Hurstpierpoint and William Mitten', www.sussexbryophytes.wordpress.com, 22 May 2019; Sue Rubinstein, 'In the Footsteps of William Mitten', www.sussexbryophytes.wordpress.com, 25 September 2018.
21 Madeline Hutchins, *Ellen Hutchins (1785–1815): Botanist of Bantry Bay* (Bantry Bay, 2019), www.ellenhutchins.com, accessed 27 November 2023; Madeline Hutchins, 'New Finds Add to the Story of Ireland's First Female Botanist, Ellen Hutchins', *Irish Heritage News*, www.irishheritagenews.ie, 1 January 2024; 'Ellen Hutchins Festival: A Festival Celebrating Botany, Botanical Art and the Beauty of Bantry Bay', www.ellenhutchins.com, accessed 27 November 2023.
22 Quoted in 'Ellen Hutchins and Ardnagashel Estate', www.ardnagashel.wordpress.com, 4 February 2016.
23 Ibid.
24 'Correspondence from 1811', www.ellenhutchins.com, accessed 27 November 2023.
25 'Her Character and Spirit', www.ellenhutchins.com, accessed 27 November 2023.
26 Ibid.
27 'Correspondence', www.ellenhutchins.com.
28 Ibid.
29 Irwin M. Brodo and Tor Tønsberg, '*Opegrapha halophila* (Opegraphaceae), a New Lichen Species from Coastal British Columbia, Canada, and Alaska, U.S.A.', *The Bryologist*, CXXII/3 (2019), pp. 457–62.
30 Elaine Ayers, 'Richard Spruce and the Trials of Victorian Bryology', *Public Domain Review*, https://publicdomainreview.org, 14 October 2015.
31 Richard Spruce, *Notes of a Botanist on the Amazon and Andes*, vol. II (London, 1908), p. 140, digital version available at www.google.com/books.
32 Ayers, 'Richard Spruce'.
33 'Richard Spruce to Daniel Hanbury', 10 February 1873, *Notes of a Botanist on the Amazon and Andes*, vol. I, ed. Alfred Russel Wallace (London, 1908), p. xxxix, available at www.biodiversitylibrary.org.
34 Harald Sack, 'Elizabeth Gertrude Britton Knight and the Study of Mosses', http://scihi.org, 9 January 2020; 'Elizabeth Gertrude Britton, 1858–1934', https://wanderwomenproject.com, accessed 27 November 2023.
35 Howard S. Conrad, 'History of the Moss Sullivant Society', *The Bryologist*, L/4 (1947), pp. 389–401.
36 Seville Flowers, 'A Visit with Dr. Grout', *The Bryologist*, L/2 (1947), pp. 208–12.

37 Matilda Knowles, 'The Maritime and Marine Lichens of Howth', *Journal of Ecology*, II/2 (1914), pp. 134–8.
38 Mary Mulvihill, 'To Matilda Knowles: A Woman's Life in Lichens Honored in Death', *Irish Times*, www.irishtimes.com, 9 October 2014.
39 Irwin M. Brodo, Sylvia Duran Sharnoff and Stephen Sharnoff, *Lichens of North America* (New Haven, CT, 2001); for images, see 'California's State Lichen: Lace Lichen (*Ramalina menziesii*)', www.californialichens.org, accessed 27 November 2023.
40 Maurice L. Zigmond, 'The Supernatural World of the Kawaiisu', in *Flowers of the Wind: Papers on Ritual, Myth, and Symbolism in California and the Southwest*, ed. Thomas C. Blackburn (Socorro, NM, 1977), pp. 59–65, available at www.vredenburgh.org.
41 Sabrina Imbler, 'The Unsung Heroine of Lichenology', *JSTOR Daily*, https://daily.jstor.org, 26 September 2020.
42 John Walter, 'Living on the Edge', *Wombat Forestcare Newsletter*, 25 (September 2013), available at www.wombatforestcare.org.au.
43 'Dr Rebecca Yahr, Royal Botanic Garden Edinburgh', https://stories.rbge.org.uk, accessed 27 November 2023.
44 Susie Rushton, 'Rebecca Yahr: Lichenology 101', *The Gentlewoman*, 25 (2022), available at https://thegentlewoman.co.uk.
45 For other lichen twig images and stories, see Anthony Speca (@specanatura, 13 November 2023), Donna Rainey (@donnarainey4, 26 February 2024), Meg Madden (@spore_stories, 20 January 2022), Cormac's Coast (@cormac_mcginley, 31 October 2022).
46 Sue Rubenstein, 'Disco Down', www.sussexbryophytes.wordpress.com, 25 November 2022.
47 David Newman, 'By the Banks of the Teise at Lamberhurst', www.sussexbryophytes.wordpress.com, 23 January 2023.
48 Brad Scott, 'Zen and the Art of Bryological Recording', www.sussexbryophytes.wordpress.com, 10 March 2021.
49 Brad Scott, 'Searching for Mosses in Early Eighteenth Century Sussex', *Sussex Botanical Recording Society Newsletter*, 92 (January 2021), available at www.sussexflora.org.uk.
50 Scott, 'Zen'.
51 Mark Frauenfelder, 'Lichens Never Looked More Beautiful Than They Do in This Short Film about a Curator of Lichen at the University of California', www.boingboing.net, 25 July 2017.
52 Daniel Gumbiner, 'The Ex-Anarchist Construction Worker Who Became a World-Renowned Scientist', *The Atlantic*, www.theatlantic.com, 19 May 2016; see also podcasts 'Kerry Knudsen on Lichen and Life after Capitalism', www.forthewild.world, 3 November 2021, and 'Kerry Knudsen –The Magic of Lichen', https://podcast.naturesarchive.com, 12 April 2021.
53 Frauenfelder, 'Lichens'.
54 Kay Harel, *Darwin's Love of Life: A Singular Case of Biophilia* (New York, 2022), p. 10.

55 Kenneth Kellman, 'The Role of the Amateur in Bryology: Tales of an Amateur Bryologist', *Fremontia*, XXXI/3 (2003), pp. 21–35.

8 #moss#lichen

1 John W. Thieret, 'Bryophytes as Economic Plants', *Economic Botany*, X/1 (1956), pp. 75–91.
2 Stephen Harrod Buhner, *The Lost Language of Plants: The Ecological Importance of Plant Medicines to Life on Earth* (Hartford, VT, 2002), p. ix.
3 'Lichens', *The Edinburgh Encyclopaedia* (Philadelphia, PA, 1932), p. 36, digital version available at www.google.com/books.
4 Riccardo Motti, Anna Di Palma and Bruna de Falco, 'Bryophytes Used in Folk Medicine: An Ethnobotanical Overview', *Horticulturae*, IX/2 (2023), p. 137; Atakan Benek, Kerem Kanli and Ergin Murat Altuner, 'Traditional Medicinal Uses of Mosses', *Anatolian Bryology*, VIII/1 (2022), pp. 57–65.
5 'Tudor Remedies: Moss', www.themerrytudor.com, 5 March 2021.
6 Eric S. J. Harris, 'Traditional Uses and Folk Classification of Bryophytes', *The Bryologist*, CXI/2 (2008), pp, 169–217, at p. 175.
7 Jacek Drobnik and Adam Stebel, '*Brachythecium rutabulum*, a Neglected Medicinal Moss', *Human Ecology*, XLVI/1 (2018), pp. 133–41.
8 Peter Ayres, 'Wound Dressing in World War I – the Kindly *Sphagnum* Moss', *Field Bryology*, 110 (2013), pp. 27–34.
9 J. W. Hotson, 'Sphagnum as a Surgical Dressing', *Science*, XLVIII/1235 (1918), available at www.digitalcollections.lib.washington.edu.
10 Drobnik and Stebel, '*Brachythecium rutabulum*'; T. Stalheim et al., 'Sphagnum – a Pectin-Like Polymer Isolated from Sphagnum Moss', *Journal of Applied Microbiology*, 106 (2009), pp. 967–76.
11 Terence J. Painter, 'Concerning the Wound-Healing Properties of *Sphagnum* Holocellulose: The Maillard Reaction in Pharmacology', *Journal of Ethnopharmacology*, LXXXVIII/2–3 (2003), pp. 145–8.
12 M. R. González-Tejero et al., 'Three Lichens Used in Popular Medicine in Eastern Andalucia (Spain)', *Economic Botany*, XLIX/1 (1995), pp. 96–8.
13 Michael E. Mitchell, 'Contentious Cures: The Rise and Decline of Lichens as European Materia Medica', *Pharmacy in History*, LVII/3–4 (2015), p. 57.
14 Ibid., p. 60.
15 Mustafa Yavuz and Gülşah Çobanoğlu, 'Ethnological Uses and Etymology of the Word Usnea in Ebubekir Razi's "Liber Almansoris"', *British Lichen Society Bulletin*, 106 (2010), pp. 3–12.
16 Tania Aebi, 'The Outside Story: Old Man's Beard', www.northernwoodlands.org, 11 June 2006; 'Fact Sheet for *Usnea longissima*', www.blm.gov, 1 December 1997.
17 P. Modenesi, 'Skull Lichens: A Curious Chapter in the History of Phytotherapy', *Fitotherapia*, LXXX/3 (2009), pp. 145–8.
18 'Skulls for Sale: English Conquest and Cannibal Medicines', *History Ireland*, XIX/3 (May/June 2011), available at www.historyireland.com.

19 Mitchell, 'Contentious Cures', p. 61.

20 '24 Eye-Catching Natural Lichen Dye Colour Drawings', www.pictureboxblue.com, accessed 28 November 2023; 'What Is a Lichen?', https://libguides.nybg.org, accessed 28 November 2023.

21 Ingvar Svanberg and Sabira Ståhlberg, 'Killing Wolves with Lichens: Wolf Lichen, *Letharia vulpina* (L.) Hue, in Scandinavian Folk Biology', *Swedish Dialects and Folk Traditions* (2017), pp. 173–87.

22 George A. Llano, 'Economic Uses of Lichens', *Economic Botany*, II/1 (1948), p. 36.

23 Ibid.

24 Dominique Cardon, *Natural Dyes: Sources, Tradition, Technology and Science* (London, 2007), p. 500.

25 Ibid., p. 249.

26 Ibid., p. 173.

27 Annette Kok, 'A Short History of the Orchil Dyes', *The Lichenologist*, III/2 (1966), p. 249.

28 Chris Cooksey, 'Tyrian Purple: The First Four Thousand Years', *Science Progress*, XCVI/2 (2013), p. 171.

29 Cardon, *Natural Dyes*, p. 501.

30 Ibid., p. 502.

31 Begoña Aguirre-Hudson, Isabella Whitworth and Brian M. Spooner, 'J. M. Despréaux' Lichens from the Canary Islands and West Africa: An Account of a 19th-Century Collection Found in an English Archive', *Botanical Journal of the Linnean Society of London*, CLXVI/2 (2011), pp. 185–211.

32 Cardon, *Natural Dyes*, p. 500.

33 Kok, 'A Short History', pp. 254, 259.

34 Ibid., p. 263.

35 Isabella Whitworth and Zvi C. Koren, 'Orchil and Tyrian Purple: Two Centuries of Bedfords from Leeds', *Ambix*, LXIII/3 (2016), pp. 244–67.

36 Isabella Whitworth, poster presentation: *A Leeds Company and the C19th Orchil Trade. Expanding Trade: Diminishing Stocks*, International Symposium and Exhibition on Natural Dyes (ISEND) (La Rochelle, France, 2011).

37 See Isabella Whitworth's blog posts, 'Talking Orchil', 'Talking Purple' and 'A Harris Way of Life', www.isabellawhitworth.com, accessed 22 January 2024.

38 Karen Diadick Casselman, *Lichen Dyes: The New Source Book* (New York, 2011), p. 1.

39 Eileen M. Bolton, *Lichens for Vegetable Dyeing* [1960], ed. Karen Leigh Casselman and Julia Bolton Holloway, 2nd edn (McMinnville, OR, 1991), digital version available at www.google.com/books.

40 Ibid., p. 8.

41 George Schenk, *Moss Gardening: Including Lichens, Liverworts, and Other Miniatures* (Portland, OR, 1997), p. 31.

42 Keir Davidson, *A Zen Life in Nature: Musō Soseki in His Garden* (Ann Arbor, MI, 2007), pp. 88, 263–4.

43 Schenk, *Moss Gardening*, p. 31.

44 François Berthier, *Reading Zen in the Rocks: The Japanese Dry Landscape* (Chicago, IL, 2000), pp. 131–2.
45 See www.sakonnetgarden.net, accessed 28 November 2023.
46 Quoted in Michael Tortorello, 'Gathering Moss', *New York Times*, www.nytimes.com, 2 April 2014.
47 Alison Pouliot, 'Lichens in the Garden – and Attic', *Australian Garden History*, XXVIII/3 (2017), p. 18.
48 Florike Egmond, *Eye for Detail: Images of Plants and Animals in Art and Science 1500–1630* (London, 2017), pp. 224–30.
49 Ibid., p. 230; see also Stefan Hanß, 'From Observation to Inspection: Florike Egmond on Microscopic Records in Early Modern Visual Albums of Nature', www.sites.manchester.ac.uk, 24 June 2020.
50 Lizzie Harper, 'Introduction to Lichens', www.lizzieharper.co.uk, 23 April 2013.
51 Lizzie Harper, 'Botanical Illustration: Step-by-Step: Spagnum Moss', www.lizzieharper.co.uk, 23 December 2016.
52 'Lichens Create One-of-a-Kind Nature Art', *Greenability*, www.greenabilitymagazine.com, 5 August 2015.
53 Donna Haraway, 'Tentacular Thinking: Anthropocene, Capitalocene, Chthulucene', *e-flux Journal*, 75 (2016), available at www.e-flux.com.
54 'We Are All Lichens', 7 October–15 December 2022, available at www.musee-rochechouart.com.
55 Nastassja Noell, 'Lichen Artists and Artistic Lichenologists: Becoming What We Attend To', *Evansia*, XL/3 (2023), pp. 96–109.
56 Ibid., p. 107; see also www.beinglichen.org.
57 Archibald MacLaren, 'The Moss-Woman', in *The Fairy Family: A Series of Ballads and Metrical Tales Illustrating the Fairy Mythology of Europe* (London, 1857), pp. 29–32, digital version available at www.googlebooks.com.
58 See www.kimsimonsson.com; Eric David, 'Kim Simonsson's Posse of Moss Children Roam a Nature-Stricken Wasteland', www.yatzer.com, 7 December 2022.
59 Tomasz Wesołowski and Sylwia Wierzcholska, 'Tits as Bryologists: Patterns of Bryophyte Use in Nests of Three Species Cohabiting a Primeval Forest', *Journal of Ornithology*, 159 (2018), pp. 733–45.
60 Francisco E. Fontúrbel et al., 'Mamma Knows Best: Why a Generalist Hummingbird Selects the Less Abundant Moss for Nest Building', *Ecology*, CI/7 (2020), e03045.
61 Francisco E. Fontúrbel et al., 'Cryptic Interactions Revisited from Ecological Networks: Mosses as a Key Link between Trees and Hummingbirds', *Functional Ecology*, XXXV/1 (2021), pp. 226–38.
62 Vilmos Molnár, 'Newly Hatched Golden Plovers Are So Well-Disguised, It's Nearly Impossible to Distinguish Them from Moss', www.earthlymission.com, 2 March 2022; 'Golden Plover Chicks – Masters of Disguise', thedefiantforest.com, 12 January 2020.
63 Gary R. Graves and Manuela Dal Forno, 'Persistence of Transported Lichen at a Hummingbird Nest Site', *Northeastern Naturalist*, XXV/4 (2018), pp. 656–61.

64 Douglas P. Kibbe, 'Northern Parula (*Parula americana*)', www.vtecostudies.org, accessed 28 November 2023.
65 'Bombus muscorum (Moss Carder Bee)', www.flickr.com; 'Bombus muscorum (Linnaeus, 1758)', www.bwars.com, both accessed 28 November 2023.
66 'The Moss Carder Bee', in *The Parlour Menagerie* (London, 1882), p. 220.
67 D. K. Upreti and Sanjeeva Nayaka, 'Need for Creation of Lichen Gardens and Sanctuaries in India', *Current Science*, XCIV/8 (2008), pp. 976–8; 'India's First Cryptogamic Garden Opens in Dehradun', *Indian Express*, www.indianexpress.com, 11 July 2021.
68 Ryszard Ochyra, Halina Bednarek Ochyra and Ronald I. Lewis Smith, '*Schistidium deceptionense*, a New Moss Species from the South Shetland Islands, Antarctica', *The Bryologist*, CVI/4 (2003), pp. 569–74.
69 Victoria Gill, 'Climate Change Kills Antarctica's Ancient Moss Beds', *BBC News*, www.bbc.com, 24 September 2018; Robin McKie, '"Simply Mind-Boggling": World Record Temperature Jump in Antarctic Raises Fears of Catastrophe',*The Guardian*, www.theguardian.com, 10 April 2024.
70 Yelena I. Kosovich-Anderson and William Weber, 'Mosses of Wyoming's Beartooth Plateau: New Noteworthy Records for the Rocky Mountain Region', *Phytoneuron*, 58 (2011), pp. 1–10.
71 Sara Hudston, 'Country Diary: Goldeneye Lichen's Quiet Resurgence', *The Guardian*, www.theguardian.com, 16 March 2019.
72 Andrea Woodward, 'Rock Gnome Lichen (*Gymnoderma lineare*) Monitoring Assessment, Southern Appalachian Mountains, 1983–2008', Open File Report 2021-1011, U.S. Geological Survey, www.pubs.usgs.gov, accessed 28 November 2023.
73 'Cornish Path-Moss: Creating Habitat for This Ultra-Rare Plant to Flourish', www.naturebftb.co.uk, accessed 28 November 2023.
74 Alejandra Tauro et al., 'Field Environmental Philosophy: A Biocultural Ethic Approach to Education and Ecotourism for Sustainability', *Sustainability*, XIII/8 (2021), available at www.mdpi.com.
75 Bernard Goffinet et al., *Miniature Forests of Cape Horn: Ecotourism with a Hand Lens* (Denton, TX, 2012).

Further Reading

Ahmadjian, Vernon, and Mason E. Hale, eds, *The Lichens* (New York and London, 1973)

Allen, Jessica L., and James C. Lendemer, *Urban Lichens: A Field Guide for Northeastern North America* (New Haven, CT, and London, 2021)

Atherton, Ian, Sam Bosanquet and Mark Lawley, eds, *Mosses and Liverworts of Britain and Ireland: A Field Guide* (Plymouth, 2010)

Bell, Neil, *The Hidden World of Mosses* (Edinburgh, 2023)

Bland, John, *Forests of Lilliput: The Realm of Mosses and Lichens* (Englewood Cliffs, NJ, 1971)

Brodo, Irwin M., Sylvia Duran Sharnoff and Stephen Sharnoff, *Lichens of North America* (New Haven, CT, and London, 2001)

Gilbert, Oliver, *Lichens* (London, 2000)

—, *The Lichen Hunters* (Sussex, 2004)

Glime, Janice M., *Bryophyte Ecology* (Houghton, MI, 2017), ebook found at www.digitalcommons.mtu.edu/bryophyte-ecology/

Goffinet, Bernard, et al., *Miniature Forests of Cape Horn: Ecotourism with a Hand Lens* (Denton, TX, 2012)

Grout, A. J., *Mosses with Hand-Lens and Microscope* [1903] (Ashton, MD, 1965)

Hinds, James W., and Patricia L. Hinds, *The Macrolichens of New England* (New York, 2007)

Jenkins, Jerry, *Mosses of the Northern Forest: A Photographic Guide* (Ithaca, NY, 2020)

Kimmerer, Robin Wall, *Gathering Moss: A Natural and Cultural History of Mosses* (Corvallis, OR, 2003)

Lücking, Robert, and Toby Spribille, *The Lives of Lichens: A Natural History* (Princeton, NJ, 2024)

Lüth, Michael, *Mosses of Europe: A Photographic Flora* (www.milueth.de)

McCune, Bruce, and Linda Geiser, *Macrolichens of the Pacific Northwest* (Corvallis, OR, 2009)

McKnight, Karl B., et al., *Common Mosses of the Northeast and Appalachians* (Princeton, NJ, and Oxford, 2013)

McMullin, Troy, and Frances Anderson, *Common Lichens of the Northeastern North America: A Field Guide* (New York, 2014)

Malcolm, Bill, and Nancy Malcolm, *The Forest Carpet: New Zealand's Little-Noticed Forest Plants – Mosses, Lichens, Liverworts, Hornworts, Fork-ferns and Lycopods* (Nelson, New Zealand, 1989)
—, and —, *Mosses and Other Bryophytes: An Illustrated Glossary* (Nelson, New Zealand, 2000)
Nash, Thomas H. III, ed., *Lichen Biology*, 2nd edn (Cambridge, 2008)
Nordström, Ulrica, *Moss – From Forest to Garden: A Guide to the Hidden World of Moss* (Woodstock, VT, 2019)
Palmer, A. Laurie, *The Lichen Museum* (Minneapolis, MN, and London, 2023)
Porley, Ron, and Nick Hodgetts, *Mosses and Liverworts* (London, 2005)
Proulx, Annie, *Fen, Bog and Swamp: A Short History of Peatland Destruction and Its Role in the Climate Crisis* (New York, 2022)
Purvis, William, *Lichens* (London, 2000)
Rydin, Håkan, and John K. Jeglum, with Aljosja Hooijer, *The Biology of Peatlands* (Oxford, 2006)
Sheldrake, Merlin, *Entangled Life: How Fungi Make Our Worlds, Change Our Minds and Shape Our Futures* (New York, 2020)
Smith, A.J.E., ed., *Bryophyte Ecology* (London, 1982)
Struzik, Edward, *Swamplands: Tundra Beavers, Quaking Bogs, and the Improbable World of Peatlands* (Washington, DC, 2021)
Tripp, Erin A., and James C. Lendemer, *Field Guide to the Lichens of Great Smoky Mountains National Park* (Knoxville, TN, 2020)
Vanderpoorten, Alain, and Bernard Goffinet, *Introduction to Bryophytes* (Cambridge, 2010)
Walewski, Joe, *Lichens of the North Woods: A Field Guide to 111 Northern Lichens* (Duluth, MN, 2007)
Williams, Sue Alix, *Ecological Guide to the Mosses and Common Liverworts of the Northeast* (Ithaca, NY, and London, 2023)
Wirth, Volkmar, *Lichens of the Namib Desert: A Guide to Their Identification* (Göttingen, 2010)
Zonca, Vincent, *Lichens*, trans. Jody Gladding (Cambridge and Hoboken, NJ, 2023)

Associations and Websites

American Bryological and Lichenological Society
www.abls.org

British Bryological Society
www.britishbryologicalsociety.org.uk

British Lichen Society
www.britishlichensociety.org.uk

Centre for Australian National Biodiversity Research
www.cpbr.gov.au

Consortium of Lichen Herbaria
https://lichenportal.org

Global Consortium of Bryophytes and Lichens
https://globaltcn.utk.edu

Inaturalist
www.inaturalist.org

International Association for Lichenology
www.ial-lichenology.org

Lichen Identification
Facebook group

Lichens Connecting People
Facebook group

Lichens and People, a bibliographic database
compiled by Sylvia Duran Sharnoff
www.sharnoffphotos.com

Michael Lüth: Photographs of Mosses and Liverworts
www.milueth.de

Moss Appreciation Society
Facebook group

Mushroom and Lichen Dyers United
Facebook group

New Zealand Plants: Mosses
www.nzplants.auckland.ac.nz

Ohio Moss and Lichen Association
www.ohiomosslichen.org

Sussex Bryophytes
www.sussexbryophytes.wordpress.com

Ways of Enlichenment
www.waysofenlichenment.net

Acknowledgements

It has been a privilege to write about mosses and lichens. These ancient organisms become our constant companions once we are attuned to them. I notice after rain that countless minute grey-green lichens inhabit a table on the back terrace and that tiny feathers of golden-green mosses have colonized the pitted surface of an unusual rock by the back door. Exemplary life forms, they improve the health of the planet wherever they are.

A work of nonfiction is usually built upon the work of others. I owe tremendous thanks to my sources, archived in the endnotes, and apologize for any errors of interpretation and notation. I offer special thanks to Alex Ciobanu, Emma Devlin, Michael Leaman and others at Reaktion who work together to produce beautiful books; bryologists and lichenologists Des Callaghan, Sean Edwards, Janice Glime (Michigan Technological University), Bernard Goffinet (University of Connecticut), Trevor Goward, Martin Grube (University of Graz, Austria), Claire Halpin, Rory Hodd, Jason Hollinger, Štěpán Koval, Jon Lendemer (New York Botanical Garden), Michael Lüth, Sharon Robinson (University of Woolongong), Gordon Rothero, Jonathan Shaw (Duke University) and Toby Spribille (University of Alberta); my friends Jane Bain (photographic specialist), Kate Bloodgood (editorial specialist), Frances Fawcett (scientific illustrator), Tina Mosetis (positivity coach); and all those who inspired me with their photos from farflung places, screenshots from social media and in other ways: David Fernandez, David Heldring (Sweden), Daniel C. Levine (Iceland; Artistic Director, The Ridgefield Playhouse), Lena Rice and Douglas Fernandez (Italy), Liz Shelbred (Olympic National Park), Charlotte Whalen (social media), Jack Whalen and Lindsay and Mike Zausmer (Norway).

Many others have contributed: Stephen Atkinson (Natural History Museum Images), Seathra Bell (www.stravaiginyarnco.com), Rosanna van den Bogaerde (Picture Library, Ashmolean Museum), Jan Brett (children's book author and illustrator), Richard Broughton (ecologist), Laura Bryer (Chetham's Library), Bryonet-L list serve, Debra Cobon (Communications Coordinator and Office Manager, Salt Spring Conservancy), Rachel Dickinson (reader), Richard Droker, David Dyer (Natural History Curator, Ohio History Connection), Florike Egmond (University of Leiden), Tamara Fulcher (Communications and Marketing

Lead Officer, Galloway and Southern Ayrshire UNESCO Biosphere), Malcolm Haddow (SWSEIC Support Officer, South West Scotland Information Centre), Stephen Harris (Oxford University Herbaria), Karolina Heyduk (University of Connecticut), Zoe Hill (Houghton Library, Harvard University), Christa Hofman (geologist/paleobotanist, University of Vienna), Christa Hofman (Head of Conservation, Austrian National Library), Daniel Hotz, Madeline Hutchins (www.ellenhutchins.com), Colm Malone (Clara Bog Nature Reserve), Serena Marner (Oxford University Herbaria), Melissa Minty (Penguin Random House), Allee Monheim (University of Washington), Andy Murray (www.chaosofdelight.org), Dan Nelson (www.10000thingsofthepnw.com), Lisa Oberg (Interim Director/History of Science and Medicine Curator, Special Collections, University of Washington Libraries), Emma Perry (University of Maine), Hellen Pethers (Research Services Librarian, Library and Archives, Natural History Museum), Natalie and Cory Pinter, Janine Quinn (Clara Bog Nature Reserve), Erin Clements Rushing (Outreach Librarian, Smithsonian Librairies and Archives), Ute Schmidthaler (Austrian National Library), Damian Shields (www.damianshields.photoshelter.com), Casey Soules (Odyssey Bookstore), Patrik Stridvall and Anita Stridvall (www.stridvall.se), Mary Sullivan (Penguin Random House), Sarah Taylor, Chris D. Thomas (John Cowper Powys Society), Edna Weber, Caroline Winschel (Director of Development and Communications, Bartram's Garden) and Isabella Whitworth (www.isabellawhitworth.com).

Photo Acknowledgements

The author and publishers wish to express their thanks to the sources listed below for illustrative material and/or permission to reproduce it. Some locations of works are also given below, in the interest of brevity:

Alamy Stock Photo: pp. 23 (Premaphotos), 97 (Simon Montgomery/robertharding), 99 (Adam Burton), 104 (Stephane Bowker), 106 (Adolf Martens/Panther Media GmbH); © Ashmolean Museum, University of Oxford, photo Ellie Atkins: p. 24; photos Jane Bain: pp. 51 (*top*), 65, 140; photo Ryan Batten: p. 71; Biblioteka Jagiellońska, Kraków (A 18): p. 204; © Jan Brett 2012, courtesy Penguin Random House Youth: p. 139; British Library, London (Add MS 22332, fol. 71r): p. 9; photos Des Callaghan: pp. 40–41, 117 (CC BY-SA 4.0); courtesy Carl A. Kroch Library, Division of Rare and Manuscript Collections, Cornell University, Ithaca, NY (photos Elizabeth Lawson): pp. 25, 27, 30; courtesy Clara Bog Nature Reserve: p. 123; Deutsches Museum, Munich: p. 33; photos Richard Droker: pp. 8, 114; courtesy Sean Edwards: p. 168; © F. Warne & Co. 1917, 1987, courtesy Penguin Random House Children's: p. 145; courtesy Frances Fawcett: pp. 39, 81; Flickr: pp. 12, 69 and 79 (photos Jason Hollinger, CC BY 2.0), 111 (photo Rob Oo, CC BY 2.0), 177 (photo Björn S, CC BY-SA 2.0), 184 and 212 (photos Jason Hollinger, CC BY 2.0); courtesy Gallway and Southern Ayrshire Biosphere, photo Malcolm Haddow: p. 122; photo Martin Grube, University of Graz: p. 84; photos Claire Halpin: pp. 20, 125; courtesy Stephen Harris, Oxford University Herbaria: p. 14; from Wilhelm Hofmeister and Frederick Currey, trans., *On the Germination, Development, and Fructification of the Higher Cryptogamia . . .* (London, 1862), photo Noranda Earth Sciences Library, University of Toronto: p. 34; Houghton Library, Harvard University, Cambridge, MA: p. 142; iNaturalist Canada: p. 182 (photo Connor, CC BY-NC 4.0); photo Graham Jones: p. 155; Kew Gardens, Richmond, London: p. 172; photos Štěpán Koval: pp. 28, 58; photos Elizabeth Lawson: cover, pp. 10, 13, 50, 74, 76 (*left* and *right*), 88, 89, 102, 107, 132, 156, 164, 180; photos Daniel C. Levine: pp. 6, 62 (*top*), 100; Library of Congress, Washington, DC: pp. 186, 193; The LuEsther T. Mertz Library, New York Botanical Garden: p. 175; photos Michael Lüth: pp. 44, 48, 54–5, 59, 188; from Donald A. Mackenzie, *Wonder Tales from Scottish Myth and Legend* (London, Glasgow and Mumbai, 1917), photo New

York Public Library: p. 148; photo Seán Maskey, courtesy Ellen Hutchins Festival (Correspondence of Dawson Turner, Wren Library, Trinity College, Cambridge, reproduced with the kind permission of the Master and Fellows, Trinity College, Cambridge): p. 171; photo Andy Murray, www.chaosofdelight.org: p. 51 (*bottom*); National Archives at College Park, MD: p. 191; Nature Picture Library: pp. 91 (© Robert Thompson/naturepl.com), 208 (© Pete Oxford/naturepl.com); photo Dan Nelson, www.10000thingsofthepnw.com: p. 90; Ohio History Connection, Columbus: p. 163; Österreichische Nationalbibliothek, Vienna (Cod. Theol. gr. 31, fol. 1r), photo © ÖNB: p. 197; photo © Peargrin, courtesy Leigh Faden: p. 209; photo Natalie Pinter: p. 66; courtesy the Powys Society: p. 151; photo Sharon Robinson: pp. 108–9; photo Gordon Rothero: p. 128; photo Fred Rumsey: p. 129; from W. Ph. Schimper, *Versuch einer Entwickelungs-Geschichte der Torfmoose (Sphagnum)* (Stuttgart, 1858), photos Peter H. Raven Library, Missouri Botanical Garden, St Louis: pp. 11, 126, 135; Science Photo Library: p. 115 (Steve Gschmeissner); photo Damian Shields, www.damianshields.photoshelter.com: p. 94; Shutterstock.com: pp. 45 (Tina Horne Photo), 63 (Jeff Holcombe), 68 (Maple Ferryman), 113 (Matauw); Smithsonian Libraries and Archives, Washington, DC: p. 18; photo Leif and Anita Stridvall, www.stridvall.se: p. 178; photo Sarah Taylor, George Safford Torrey Herbarium, University of Connecticut, Storrs: p. 62 (*bottom*); © The Trustees of the Natural History Museum (WP/2/1/24): p. 166; photo Edna Weber: p. 157; Lew Welch, 'Springtime in the Rockies, Lichen' from *Ring of Bone: Collected Poems 1950–1971*, copyright © 1979 by Donald Allen, Executor of the Estate of Lew Welch, reprinted with the permission of The Permissions Company, LLC on behalf of City Lights Books, www.citylights.com: p. 154; Wellcome Collection, London (CC BY 4.0): p. 36; courtesy Isabella Whitworth, www.isabellawhitworth.com: p. 200; Wikimedia Commons: pp. 70 (photo Tigerente, CC BY-SA 3.0), 92 (photo T.Voekler, CC BY-SA 3.0), 105 (photo Olga Ernst, CC BY-SA 4.0), 119 (photo Bernd Haynold, CC BY-SA 3.0), 133 (photo Alan Rockefeller, CC BY-SA 4.0), 149 (photo B.gliwa, CC BY-SA 2.5), 196 (photo Norbert Nagel, CC BY-SA 3.0), 201 (photo 江戸村のとくぞう, CC BY-SA 4.0), 210 (photo Panoramedia, CC BY-SA 3.0); Yale Center for British Art, New Haven, CT: pp. 121, 198.

Index

Page numbers in *italics* refer to illustrations

Acarospora socialis 183, *184*
Accademia dei Lincei (Academy of the Lynx-Eyed) 22–3, 203
Acharius, Erik 68, 87, 165
alternation of generations 38, *39*
Antarctica 108–10, *108–9*, *111*, 115, 211–12
antlered jellyskin lichen *8*, *90*, *114*
apple-pelt lichen *69*
Archidium phascoides 34
Arctic 108, 132, 207
Arora, Nikita, 'The Many Meanings of Moss' 159–60
asexual reproduction *39*, 45–6, 57, *81*

Barnes, William, 'Moss' *140*
Bartram, John 15, 17–20, *18*, 22
Bartramia pomiformis 20
Bary, Heinrich Anton de 71
Bay of Fires, Tasmania *104*
beard lichen *92*
Bell, Neil, *The Hidden World of Mosses* 49
Berthier, François, *Reading Zen in the Rocks* 201
biocrust 9, 103
biophilia 162, 184
bird nest material 206–9
Birr Castle statue *157*
Bishop, Elizabeth, 'The Shampoo' 158
black-fruited stink moss 59
blushing bog-moss *129*
bog (peat bog) 57, *59*, 117–21
Clara Bog, Ireland *123*
Jam Pond, New York *132*
Bolton, Eileen, *Lichens for Vegetable Dyeing* 200
Borrer, William 68, 164–5, *164*
Bower, F. O., *The Origin of a Land Flora* 35
Brachythecium rutabulum 189
Braithwaite, Robert 127–8
Brett, Jan, *Mossy 139*
bright cobblestone lichen *184*
bristle moss *44*
Britton, Elizabeth Gertrude Knight 168, 175–7, *175*
brocade moss *63*
broom forkmoss *45*
brown mosses 131
brown-eyed sunshine *79*
Bryoria sp. 77
Bryson, Bill, *A Short History of Nearly Everything* 160
Bryum argenteum 61–2, 110, 113
Buddle, Adam 16
Buellia peregrina 104
bugs-on-a-stick moss 38, 56, *58*
Buhner, Stephen Harrod, *The Lost Language of Plants* 187
Buxbaum, Johann Christian 56
Buxbaumia aphylla 38, 56, *58*, 176
Buxbaumia viridis 56–7
Buxton, Richard 20–21

Callaghan, Des 42, 49
Calliergon giganteum 132
Calopaca elegantissima *106*
Calopadia puiggarii 82
Cardon, Dominique, *Natural Dyes: Sources, Traditions, Technology and Science* 195–9
Carson, Rachel, *The Sea around Us* 7
Casselman, Karen Diadick, *Lichen Dyes: The New Source Book* 199
Cathcart, Charles W. 190
Ceratodon purpureus 38, 49, 61, *109*, 113
Cetraria islandica 191
Chesteron, G. K. 144
Church, Arthur Harry, *Thalassiophyta and the Subaerial Transmigration* 32
Cibo, Gherardo *9*
Cinclidium stygium 53
Cladina evansii *66*
Cladonia sp. 93, 113
Cladonia cristatella *102*
Cladonia pyxidata 191–2
Cladonia uncialis 100
Collema sp. 71
Collinson, Peter 19
common apple-moss *20*
common greenshield lichen *13*
common toadskin lichen *76*
conservation 211–13
Cornish path-moss 213
Crombie, Reverend James M. 73–5, 93
cruet collar moss 57
Cruikshank, Isaac, *A Will o' the Wisp, or John Bull in a Bog* *121*
Crum, Howard, *A Focus on Peat and Peatlands* 117, 131, 134
Cryptogamia 8, *34*
cryptogamic carpet 9
cryptogams 15–16, 22–4, 33
curious/curiosity 15, 17, 19, 53, 56, 159, 162, 183–5
cyanobacteria 64, *70*, 72, 86, 88–9, 103

dagoda dome, Sri Lanka *97*
Davis, Lydia, 'A Person Asked Me about Lichens' 158–9
delicate fern moss 64
desert 9, 88, 101, 117, 158
  fog deserts 103
  Namib 103–8, *105*, *106*
  North America 93, 183, *184*
  polar desert 108
  wet desert 117
desiccation tolerance (DT) 96–8
Dibaeis baeomyces *68*
Dicranum scoparium *45*, 139
Dillenius, Johannes 14, 16–19, 25
Diston, Alfred, *Gomero, Archilla Gatherer* *198*
dung mosses 57–59

earth moss *34*
Edwards, Sean R. 139, *168*
Egmond, Florike, *Eye for Detail: Images of Plants and Animals in Art and Science, 1500–1630* 203
electrified cat's tail moss 64
elegant sunburst lichen *212*
Encalypta ciliata *25*, *27*, *28*
Entosthodon templetoni 22
extremophile 95, 112

feather moss *63*, 64, *76*, 131, 189
Finland 55, *59*, 146
fire dot lichen *106*
fire extinguisher moss *28*
fire moss 38, 49, 61
Fisher, Kirsten 105
Fissidens parkii *168*
Flavoparmelia caperata *13*
Frank, Albert Bernhard 76
fringed hoar moss *23*

Gander, Forrest, *Twice Alive: An Ecology of Intimacies* 152–3, 155
Gilbert, Elizabeth, *The Signature of All Things* 138–9
goblin gold 42
Goethe, Johann Wolfgang von 29–30
golden eye lichen 212–13

Goodenough, Ursula 83, 85
Goward, Trevor 80, 85, 87, 91, 93
Great Goldilocks moss *14*
Grieve, Hilda 15
Grimmia apocarpa 110
Grout, A. J. 53, 175
Grube, Martin 78, 82, *84*, 86

haircap moss 14, 49
Harel, Kay, *Darwin's Love of Life* 183–4
Harper, Lizzie 204
Hawksworth, David 78
Heaney, Seamus, 'Bogland' 123
Hearn, Sarah 204–5
Hedwig, Johann 20, 23–9, *24*, *30*, 53
  Species muscorum frondosorum *25*, *27*
Hedwigia ciliata *23*
Herron, Mick, *Slow Horses* 137
Hillman, Brenda, *Extra Hidden Life, among the Days* 155–6
Hiroshige, Utagawa, *Iwatake Gathering at Kumano in Kishū* *186*
Historia muscorum 16, 19, 25
Historia plantarum 16
Historia plantarum universalis 189
hoary fringe-moss *114*
Hobson, Edward, 20–22
  Musci Britannici 20, 22
Hofmeister, Wilhelm 32–3, *33*, *34*
Holten, Katie 151
Honegger, Rosmarie 93
Hooke, Robert 23, *36*, 37, 64
Hooker, William 21, 166, 170
Horsefield, John 21–2
Hotson, J. W. 121, 190–91
Hutchins, Ellen 169–72, *171*
Hygroamblystegium varium 47
Hypnum imponens *63*

Iceland *6*, 61, *62*, *100*, 147
Isothecium myosuroides *48*

Jansson, Tove, *The Summer Book* 146–7
Jardin des Plantes 31
jelly lichens 71
jellyskin lichens *8*, 86, *90*, *114*
Jenkins, Jerry 48, 61, 99
Johnson, Samuel 161
Jussieu, Antoine, Antoine-Laurent and Bernard de 31–2

Kellman, Kenneth, 185
Killarney National Park *91*
Kimmerer, Robin Wall 57, 138, 150
Knowles, Matilda Cullen 176
Knudsen, Kerry 183–4

Lasallia papulosa *76*
Lawley, Mark 20
lemmings 59
Lempholemma polyanthes *71*
Lepraria *84*, *164*
Letharia vulpina 154, *155*, 194–5
Lewis, Lily 42
Libri Picturati 203, *204*
lichen 30, 82, *180*
  apothecium/apothecia 76, 68, 80–82, *81*, *102*, *107*, *164*, 172
  biofilm 9, 85
  chasmolithic 104
  chlorobiont (green algae) 71–2, 85–6
  crustose 10, 76, 84, 86–7, *88*, *94*, 99, 103, *104*, 110, 112
  cyanobiont *82*, 85
  cyanolichen 102
  Cyphobasidium yeasts 78
  dyes 194–200
  epilithic 103–4
  euendolithic 104
  foam lichens 93, *178*
  foliose *74*, 87, 89, 110, 112
  fossil 10, 12
  fruticose *66*, 87, 89, 93, 110
  hydrophobins 85
  hyphae 72, 83–4
  ice nucleators *113*
  life cycle *33*, 80, *81*
  macrolichen 13, 89
  medicinal 191–3
  microbiome 78
  modular growth 83

mycobiont 80, 86, 153
novelty 87
photobiont 82–3, 153
soredia *81*, *82*, 84
spore 29, 81
thallus 70, *81*, 83–7
tripartite *70*
vagrant/erratic 101, 110
lichenization 11, 97
Lindsay, William Lauder, *A Popular History of British Lichens* 161
Linnaeus, Carl 8, 19, 22, 24, 26, 29, 31, 68, 80, 117, 119, 191–3, 210, 215
Lobaria pulmonaria *9*, 80, 82, 85, 89–92, 191
Lobaria scrobiculata *91*
Lobarion 92
Loch Scavaig, Scotland *94*
lurid cupola-moss 53
lustrous bog-moss *125*
Lüth, Michael 49, 60

Macdonald, Helen, *H Is for Hawk* 159
Macfarlane, Robert 150–51, 160
Mackenzie, Donald A., 'The Princess of Land-under-Waves' 148–9, *148*
Mackenzie, Elke 178
Mägdefrau, Karl 60
map lichen 13
Margulis, Lynn 78–9
Marilaun, Anton Kerner von 43
Martin, Annie, *The Magical World of Moss Gardening* 201–2
Mastodia tessellata 110
Mellor, Ethel 99–100
Melville, Herman, *Moby Dick* 64–5, 137
Metamorphosis of Plants, The 31
Miller, Madeline, *Circe* 146
Milne, Drew 151–2
Mishler, Brent 61
mites, armoured 112
Mitten, William 165–9, *166*
moose (*Alces americanus*) 58, 60
moss *36*, *39*, *40–41*
acrocarp *39*, *40–41*, *44*, *45*, 60–61, *62*, 64, 207
antheridium/antheridia 26, *39*, 49
archegonium/archegonia 26, 52
capsule *39*, 53, 56–7, 129–30
cushion *10*, *62*, *65*
fossil 10, 12
gametophyte 33, *39*, 42–4, 56, *62*, 102
leaf anatomy 45–7
leaf tip 26, 46, 101, 103, 105–7
life cycle 38, *39*
medicinal 188–91, *191*
modular growth 43
moss balls 111–12
operculum *39*, 53
pleurocarp 60, *63*, 64, 131, 207
protonema *39*, 42, 56
seta/setae 53, 56
species complex 44, 128
sperm 26, *27*, 49
splash cups 49, *50*
spore 38, *39*, 42, 57, 130
sporophyte 34, *39*, *49*, 56
vagrant 128
moss carder bee (*Bombus muscorum*) *210*
mourning phlegm lichen *71*
mouse-tail moss *48*

Nägeli, Carl 72
Namib/Namibia 103–5, *105*
Neckera douglasii *188*
nematodes 112–14
Noell, Nastassja 206
Nostoc 71–2, 89
nunatak 110
Nylander, William 73–5, 165

old man's beard lichen 209
Oliver, Mary, 'The Moss' 159
Opegrapha sp. 171–2, 179, 181
orange chocolate chip lichen *70*
orchil lichen dye 195–200
Orthotrichum sp. *44*, 53

Parmelia parietina 70
Parmelia saxatilis 192, 194
Peltigera sp. 191

Peltigera malacea 69
Percy, John 22
pink earth lichen *68*
Plagiomnium insigne 189
Plitt, Charles C. 68
Polytrichum *39*, 49, 179, 189
Polytrichum commune *14*, 17, 53
Polytrichum formosum 52
Polytrichum juniperinum 50
Polytrichum piliferum 114
Pope, Alexander 144–6
Porpidia albocaerulescens 12
Potter, Beatrix, *The Tale of Timmy Tiptoes* *145*, 146
Powys, John Cowper, *A Glastonbury Romance* 150, *151*
Pringle, Anne 152–3
Proctor, M.C.F. 37, 46
Pseudocrossidium hornsuchianum 181, *182*
Pullman, Philip, 'Mossycoat' 147
Pyle, Howard 18

Quignard, Pascal, *Mysterious Solidarities* 152

Racomitrium canescens 114
Racomitrium lanuginosum 61, *100*
Radcliffe, Ann 141
Ramalina 93
Ramalina farinacea 86, 176–7
Ramalina menziesii 176
Ratcliff, Marc 23
Ray, John 16
Réaumur, René-Antoine Ferchault de 15, 22
red parasol moss *59*, 60
reindeer lichens 93
Rhizocarpon geographicum 13, 87
Rhodobryum giganteum 188
Rhytidiadelphus triquetrus 64
Richards, P. W. 64
Richardson, Ralph 16–17
Roach, Margaret 203
Robinson, Marilynne, *Jack* 156–7
Roccella tinctoria 196
rock gnome lichen 213
rotifers 112
Royal Botanic Garden Edinburgh 179, 211
Rozzi, Ricardo 213–14
Rusavskia elegans 212
Ruskin, John, 'Study of Rocks and Lichens in the Glen . . .' *142*
Russell, Karen, 'The Bog Girl' 117

Saihō-ji moss temple 201–2, *201*
Sakonnet Garden, Rhode Island 202
salvage botany 200, 202
Sanders, William 82
Saunton Down, North Devon 39, *40–41*
Schimper, W. P., *A History of Development of Peat Mosses* *11*, *126*, 134, *135*, 162
Schistidium deceptionense 211
Schistostega pennata 42–3
Schwendener, Simon 8, 72–4, 83
Schwendenerism 73–5
Scott, Brad 167, 182–3
scribble lichens 171–2, 181
Scytinium palmatum *8*, *90*, *114*
Seaward, Mark R. D. 164
Secord, Anne 21
Shakespeare, William, *The Comedy of Errors* 143
Sheldrake, Merlin, *Entangled Life* 87, 93
Silver Flowe, Scotland *122*
silver moss 61–2, 110, 113
Simonsson, Kim, 'Moss People' 206
skull moss/lichen 191–3, *193*
Skye bog-moss *128*
Smith, Annie Lorrain, *Lichens* 67, 75
smoky-eye boulder lichen *12*
Solorina crocea 70
Soseki, Musō 201
sphagnan 131, 190–91
sphagnorubins *149*
sphagnum 11, 60, 114, 117–36
species complexes 128
sphagnum dressings 190–91, *191*
spore explosion 129–30
Sphagnum acutifolium *11*, *135*

Sphagnum mole 129
Sphagnum rubellum 116, 149
Sphagnum skyense 128
Sphagnum squarrosum 119
Sphagnum subnitens var. *subnitens 125*
Splachnum ampullaceum 57
Splachnum luteum 54–5, 57
Splachnum rubrum 59, 60
Splachnum sphaericum 58
Spribille, Toby 77, 82, 86
springtails (Collembola) 49, *51*, 52
Spruce, Richard 172–4, *172*
Stark, Lloyd R. 104–5
Stereocaulon 179
Stereocaulon paschale 178
Sullivant, William Starling 162–3, *163*
sundews *133*
Sussex Bryophytes 181–3
symbiogenesis 78
symbiosis 75–7, 85, 110, 205, 210
  kleptosymbiosis 86
symbiotic dust *84*
Syntrichia caninervis 103, 105–7

Taiz, Lincoln and Lee, *Flora Unveiled* 22
tangled thread moss 47
tardigrades 112–14, *115*
Tayloria octoblepharum 57
Teloschistes chrysophthalmus 212
temperate rainforest 91
terrestrialization 12, 131
Tetraplodon fuegiensis 57
Tetraplodon mnioides 59, *62*
Thieret, John W., 'Bryophytes as Economic Plants' 187
Thoreau, Henry D., *Journals* 160
Thuidium delicatulum 64
Thwaites, George Henry Kendrick 71–3, 83
Timiryavez, Kliment 32
Tolkien, J.R.R., *The Return of the King* 143–4
Trematodon ambiguus 128
Turner, Dawson 165, 169–71

Umbilicaria 75, *186*
Usnea 93, 187, 191–2, 209
Usnea antarctica 92
Usnea cavernosa 187
Usnea longissima 102, 217

Verrucariaceae 102, 110, 158, 176
Vienna Genesis, 'Fall of Man' *197*
Volmer, Stephanie 19
Vulpicida canadensis 79, 80
vulpinic acid 77, 101

Welch, Lew, 'Springtime in the Rockies, Lichen' 153–4, *154*, 159, 214
Whalen, Philip 153
Whitworth, Isabella 199–200, *200*
Williams, Sue 49
Wirth, Volkmar, *Lichens of the Namib Desert* 103
Wistman's Wood *99*
Wodehouse, P. G. 142–3
wolf lichen 155, *155*, 194–5
woolly fringe-moss 61, *100*
Wordsworth, Dorothy 141–2
Wyndham, John, *Trouble with Lichen* 153

Xanthoparmelia 98
Xanthoria elegans 115

Yahr, Rebecca 179
yellow moosedung moss *54–5*
Yong, Ed, 67